About Island Press

Island Press is the only nonprofit organization in the United States whose principal purpose is the publication of books on environmental issues and natural resource management. We provide solutions-oriented information to professionals, public officials, business and community leaders, and concerned citizens who are shaping responses to environmental problems.

In 1999, Island Press celebrates its fifteenth anniversary as the leading provider of timely and practical books that take a multidisciplinary approach to critical environmental concerns. Our growing list of titles reflects our commitment to bringing the best of an expanding body of literature to the environmental community throughout North America and the world.

Support for Island Press is provided by The Jenifer Altman Foundation, The Bullitt Foundation, The Mary Flagler Cary Charitable Trust, The Nathan Cummings Foundation, The Geraldine R. Dodge Foundation, The Charles Engelhard Foundation, The Ford Foundation, The Vira I. Heinz Endowment, The W. Alton Jones Foundation, The John D. and Catherine T. MacArthur Foundation, The Andrew W. Mellon Foundation, The Charles Stewart Mott Foundation, The Curtis and Edith Munson Foundation, The National Fish and Wildlife Foundation, The National Science Foundation, The New-Land Foundation, The David and Lucile Packard Foundation, The Pew Charitable Trusts, The Surdna Foundation, The Winslow Foundation, and individual donors.

Practical Approaches to the Conservation of Biological Diversity

Practical Approaches to the Conservation of Biological Diversity

Edited By

Richard K. Baydack,
Henry Campa III, and
Jonathan B. Haufler

ISLAND PRESS
Washington, D.C. • Covelo, California

Library of Congress Cataloging-in-Publication Data
Practical approaches to the conservation of biological diversity /
  edited by Richard K. Baydack, Henry Campa III, Jonathan B. Haufler.
      p.   cm.
    Includes bibliographical references and index.
    ISBN 1–55963–543–6 (cloth : alk. paper). — ISBN 1–55963–544–4
  (pbk. : alk. paper)
    1. Biological diversity conservation—Congresses.   I. Baydack,
  Richard K.   II. Campa, Henry.   III. Haufler, Jonathan B.
  ᴜQH75.A1P735   1999   98–25753
    333.95'16—dc21   CIP

Printed on recycled, acid-free paper

Manufactured in the United States of America
10 9 8 7 6 5 4 3 2 1

Contents

Part 3: Opportunities and Challenges

Part 4: Summary and Recommendations

Preface

Conservation of biological diversity is critical to maintaining the future of life on Earth as we know it. Conversely, the lack of such conservation will undoubtedly lead to continued elimination of life-forms, loss of genetic material, and disruption of natural processes. Loss of biological diversity will also have enormous costs that could be described in terms of traditional economic indicators, social implications, ethical considerations, aesthetic factors, and spiritual components. In short, conservation of biological diversity can be linked to the continued existence of all life on Earth, including that of humankind.

Burgeoning human populations and their associated demand on natural resources have created many of the challenges to biological diversity that we are facing today. Human impacts on species have caused numerous local extirpations that, considered together, mean an overall and significant decline in biodiversity. But as the problem has increased worldwide, so too has the attention it has received. Although the study of biodiversity first emerged in scientific disciplines about fifteen years ago, much of the interest in it was catalyzed by the Rio Conference on Environment and Development, the so-called Earth Summit. From that meeting in summer 1992, the United Nations Convention on Biological Diversity was refined and ultimately ratified on December 29, 1993. More than 150 countries are now signatories to the Convention.

The preamble to the Convention states that there is a general lack of information and knowledge regarding biological diversity, and that there is an urgent need to develop scientific, technical, and institutional capabilities in this area. Most countries have begun to respond to the Convention's challenge by developing general strategies and assessments designed to show adherence to the basic principles of the accord. Nevertheless, only a few have begun the real work of conserving biological diversity: choosing and applying specific, practical approaches that ensure conservation of biodiversity at the local or regional level.

In North America, The Wildlife Society, a professional organization of wildlife biologists from government, academic, and private institutions, formed the Biological Diversity Working Group in 1994 in response to the growing need to address various questions relating to biodiversity conservation. The Working Group became active at annual meetings of The Wildlife Society by organizing symposia that served as a forum for views on conservation of biological diversity.

This book is an expansion and refinement of the original, contemporary papers presented at the Second Annual Conference of The Wildlife Society in Portland, Oregon, from September 12 to 17, 1995. Compiled by the symposium organizers, the essays presented here address a wide range of strategies for the conservation of terrestrial biological diversity. Many of these strategies are applicable to other types of ecosystems as well. They have accumulated over time, in a variety of published works, and are continuing to evolve as research progresses.

This is the first book to compile and describe these approaches in one volume. Our goal in presenting them here is to help managers select an appropriate technique for use in their own specific situations and students to compare and contrast methods as a learning experience in various courses.

Throughout the book, we emphasize the advantages and disadvantages of different approaches, offer hints of what to watch out for in management, and discuss challenges for future efforts. The book presents the thesis that biodiversity conservation is not only workable but also practical, feasible, and necessary in our future world. We hope you will find it to offer a useful and easy-to-follow set of guidelines to approach the complex problem of conserving biological diversity.

The sixteen chapters of this book are divided into four parts.

Part 1: Conserving Biodiversity—Principles and Perspectives

Present-day "innovations" in the study of biodiversity have their roots well set in more than sixty years of debate originating with the early writings of Aldo Leopold. Therefore, Part 1 provides a survey of the essential concepts, key terminology, and historical background of biodiversity conservation to provide a context for the selection and application of restorative strategies. Current approaches are discussed at a general level. Finally, Part 1 integrates the above discussions to introduce the "real" issue for further analysis: Why do humans need to worry about conservation of biological diversity? Is it simply political rhetoric designed to attract votes from uncommitted members of the electorate? Or is it truly important to biological life-forms and their continued existence on this planet? The chapters in Part 1 explore these issues.

Part 2: Strategies for Conserving Biodiversity

Part 2 provides detailed description of recommended approaches for conserving diversity. Each approach is described from a theoretical perspective along with examples of practical case-study applications. The writers discuss the most recent innovations and "fine-tuning" that have accompanied their various approaches. Where possible, approaches are described in as many different ecosystems as possible in a wide range of locations.

Part 3: Opportunities and Challenges

This part of the book raises some key issues bearing on the application of the strategies described in Part 2—typical, real-world constraints, criticisms, and related management implications and problems that are often raised when biodiversity conservation is discussed. These include what to do if "too little" information is available, how to cope with "too much" information, how to address the age-old concern about funding, what to do if time is a factor, what to do about exotic species, and many others. The authors first describe the questions that generally arise and then offer practical alternative solutions for managers to consider.

Part 4: Summary and Recommendations

Part 4 contrasts strategies and suggests a model process for implementing them as well as identifying additional needs for improved management in the future. The section also contains a set of guidelines of "best practices" for conserving biodiversity to assist managers, administrators, and students who are currently struggling to put words into action in their respective situations. Chapter 16, "Future Innovations and Research Needs," provides researchers with a wide range of ideas that can serve as catalysts for their own research proposals. Methods for improving decision making that advances biological diversity are discussed in terms of practical significance, dollar costs, and other factors. The book concludes by proposing that it is possible and necessary to conserve biological diversity in virtually any management situation.

Acknowledgments

The editors would like to express their sincere appreciation to the many persons and organizations contributing to the development of this book. First, The Wildlife Society was instrumental in formulating the content of the book by accepting the initial concept as a conference contribution. The various chapter authors provided timely materials at the conference and in subsequent iterations. Our respective employers, the Natural Resources Institute at the University of Manitoba, the Department of Fisheries and Wildlife at Michigan State University, and the Boise Cascade Corporation provided logistical and financial support for this undertaking. Of course, our families and friends were forced to endure the long hours and late nights that this type of work entails. And, finally to the personnel at Island Press, most notably Barbara Dean and Barbara Youngblood, our thanks for your patience and encouragement throughout the process.

Richard K. Baydack
Henry Campa III
Jonathan B. Haufler

Conserving Biological Diversity—Principles and Perspectives

Part 1 provides a foundation for the other chapters in this book. By familiarizing themselves with the principles and perspectives discussed in Part 1, readers can gain a more accurate and comprehensive understanding of the concepts discussed in later chapters.

The purpose of Chapter 1 is to establish a clear linkage between what the concept of biodiversity is and what it is not, where it came from, and why it is important. The chapter serves four important functions: it defines biological diversity and other key terms; it describes the historical development of the concept, it discusses the relationships between biodiversity and related concepts, and it points out the critical need to conserve biological diversity.

Chapter 1 discusses biodiversity at all levels of biological organization, the spatial scales in which those levels occur, and the importance of ecological processes. The interrelationships between the concepts of biological diversity, ecosystem management, and conservation biology are also explored. But all of this still begs the question, Why do we need to conserve biological diversity? We discuss a variety of ecological reasons that show why the conservation of biological diversity is central to current biological management, including maintenance of ecosystem functions and processes, species interactions, genetic diversity, and direct human benefits.

Chapter 2 provides an overview of the specific technical approaches described in Part 2. The application of various techniques to landscape-level management is discussed as well as the necessary scale considerations that natural resource managers take into account as they work to conserve biological diversity. The coarse- and fine-filter approaches to landscape-level management can be viewed as two opposite ends of the spectrum in terms of maintaining biological diversity. The coarse-filter approach is based on the premise that biological diversity can be conserved by maintaining a variety of ecosystems. A common criticism of this method as a sole means to maintain biological diversity is that some species may "slip through the cracks." This problem and solutions to deal with it are discussed in detail in later chapters.

The fine-filter approach focuses on the specific habitat requirements of individual species that may not be considered by the coarse-filter method. How the requirements of individual species can be met while maximizing the conservation of biodiversity across the larger landscape is discussed in Chapter 2. Jon Haufler also describes the difficulties of dealing with practical issues, such as conflicting ownership, jurisdictions, and human-created geographic boundaries.

Like it or not, the diversity of natural resources we want to maintain does not always adhere to these types of categorizations. To effectively conserve biological diversity, natural resource managers are now considering broad-scale ecological boundaries and processes, the frequency of historic changes, and the spatial scales at which these occur. Examples of how these attributes are being used and why they need to be considered will be discussed in detail in Part 2.

Chapter 1

Setting the Context

Richard K. Baydack
and Henry Campa III

What is biodiversity? A term and a concept that did not exist ten years ago, "biodiversity," short for "biological diversity," has become "one of the most commonly used expressions in the biological sciences and . . . a household word" (Wilson 1997).

"Biodiversity" can mean different things to different people, so therefore anyone using the term needs to define it or at least imply its definition to ensure that others are aware of the specific orientation under consideration. Table 1.1 summarizes definitions that have been developed over the past fifteen years.

Numerous sources have provided detailed definitions of "biological diversity." Norse and McManus (1980) provided the first contemporary definition. Wilson and Peter (1988) brought the notion of biodiversity to the attention of a broad array of biologists and other scientists. The U.S. Forest Service (1990) offered the first working definition for field managers, and Salwasser (1990) specified its boundaries. The Keystone Center (1991) focused on different levels of biological organization that are often associated with biological diversity. The U.S. Fish and Wildlife Service (1992) attempted to amalgamate the concepts inherent in the previous approaches. Environment Canada (1996) further developed the term from a national perspective. Wilson (1997) and Lovejoy (1997) focus their discussions on the broad meaning and value of biodiversity. So is one of these definitions "correct" or at least better than others? How does one select the right definition?

At first glance you can see that all the definitions are related, although each has its own specific attribute that may be more appro-

Table 1.1. Definitions of "Biological Diversity," or "Biodiversity"

Definition	Source
"The amount of genetic variability within species and the number of species in a community of organisms."	Norse and McManus (1980)
"All life forms, with their manifold variety, that occur on earth."	Miller et al. (1985)
"The variety and variability among living organisms and the ecological complexes in which they occur."	Office of Technology Assessment (1987)
"The diversity and variability of plants, animals, microorganisms, and the ecosystems in which they occur."	Wilson and Peter (1988)
"The degree of nature's variety, including both the number and frequency of ecosystems, species, and genes in a given area."	McNeely et al. (1990)
"The variety of life and its processes."	U.S. Forest Service (1990)
"The variety of life and its processes in a given area."	Salwasser (1990)
"The variety of life and its processes. It includes the variety of living organisms, the genetic differences among them, and the communities and ecosystems in which they occur."	The Keystone Center (1991)
"The variety of and variability among living organisms and the ecological complexes of which they are part; this includes diversity within species, between species and of ecosystems."	United Nations Environment Programme (1991)
"The full range of variety and variability within and among living organisms and the ecological complexes in which they occur; encompasses ecosystems or community diversity, species diversity and genetic diversity."	U.S. Congress (1991)
"The variety of life in an area, including genetic composition, richness of species, distribution and abundance of ecosystems and communities, and the processes by which all living things interact with one another and their environment."	U.S. Fish and Wildlife Service (1992)

Definition	Source
"The total variety of life on earth."	Ryan (1992)
"Biodiversity is the variability among living organisms from all sources including, inter alia, terrestrial, marine, and other aquatic ecosystems and the ecological complexes of which they are a part."	Johnson (1993)
"Biodiversity consists of more than just the variety of species; it involves the full range of species, variation within species, biotic communities, and ecosystems in a dynamic, ever-changing process."	Noss and Cooperrider (1994)
"The total diversity and variability of living things and of the systems of which they are a part."	Heywood (1995)
"The variability among living organisms from all sources including, inter alia, terrestrial, marine and other aquatic ecosystems and the ecological complexes of which they are a part; this includes diversity within species, between species and of ecosystems."	Environment Canada (1996)
"Biodiversity is an attribute of a site or area that consists of the variety within and among biotic communities, whether influenced by humans or not, at any spatial scale from microsites and habitat patches to the entire biosphere."	Delong (1996)
"Biodiversity has to be thought of in a number of different ways—over evolutionary time, as a characteristic of natural communities, globally, and collectively."	Lovejoy (1997)
"Biodiversity is, in one sense, everything. Biodiversity is all hereditarily based variation at all levels of organization, from the genes within a single local population or species, to the species composing all or part of a local community, and finally to the communities themselves that compose the living parts of the multifarious ecosystems of the world."	Wilson (1997)

priate to a particular situation. For example, a geneticist is more likely to focus on that particular level of biological organization. A conservation biologist may select a definition appropriate to the population or ecosystem level. While each person may emphasize a different component according to his or her own perspective, what all definitions of biological diversity have in common is that the term encompasses concerns about genetics, species, populations, ecosystems, and ecological processes over evolutionary time spans.

Skiba (1994) provided a viewpoint on the issue of definitions, arguing that there is no need, and indeed no hope, for a single, standard definition. His final paragraph is most instructive.

> We have any number of workable definitions for biological diversity. Let's be sure that we have some agreement on our goals and objectives for the management, conservation, and/or enhancement of that diversity, and get moving.

What Biodiversity Is Not

Management for biodiversity is not and should not be considered a panacea to be used in all cases for "proper" natural resource management. Conversely, many objectives of natural resource management do not directly address conservation of biological diversity, although any management action undertaken to further such objectives should not significantly threaten or reduce biodiversity overall. If significant impacts to biodiversity are an outcome of those objectives, then society needs to understand the consequences of its actions and the larger threats that loss of biological diversity could lead to. In these and other situations, natural resource managers should be encouraged to express their viewpoints, supported by scientific evidence, as individuals or through professional societies, about the implications of societal decisions. It is only through an open, forward-thinking approach that these cases will be thoroughly addressed.

Finally, biodiversity should not be described as a standardized, quantifiable measure that can be readily applied from location to location. Unfortunately, there is no, and likely never will be, a universal measure for biological diversity. Many of the measurements of biological diversity are related to scale, and in any case care must be taken to identify the correct measures or objectives to be used to evaluate any specific management decision. The specific variables to be measured will very much depend on the questions being asked, the particular management decisions being made, and the levels of risk to genetics, species, populations, ecosystems, and evolutionary processes that are considered acceptable in the situation.

Historical Development of the Concept

Although Wilson (1997) suggested that the study of biological diversity can be traced as far back as Aristotle, the term was first formally defined by Norse and McManus (1980). Their definition included two related concepts—genetic diversity (the amount of genetic variability within species) and ecological diversity (the number of species in a community of organisms). Heywood (1995) has suggested that the contracted form—"biodiversity"—was apparently coined by Walter G. Rosen in 1985 for the first planning meeting of the "National Forum on BioDiversity" held in Washington, D.C., in 1986. It was at this forum that the notion was brought to the attention of a wide variety of scientists and other professionals.

Nevertheless, questions about the array or diversity of life on Earth have likely occupied the minds of humans for as long as they have inhabited the planet. Attempts to understand this diversity were aided by classification and development of the field of taxonomy from as early as the twenty-fourth century B.P. (before the present) (Heywood 1995). Reaka-Kudla et al. (1997) described the contribution to our understanding of biodiversity that was made by biblical scholars and medieval writings. Similarly, ideas about the linkages and relationships between organisms and their environments were enhanced by the studies of naturalists such as Darwin, Humboldt, Wallace, and other founders of the developing field of ecology from the eighteenth century onwards (Heywood 1995). Although some of the basic facts of heredity appear to have been known in biblical times, Mendel's experiments in the nineteenth century ushered in the field of genetics, which again has contributed to a more complete understanding of biodiversity. More recently, population biology and its emphasis on quantitative modeling have similarly expanded our understanding of the variety of life on Earth. In fact, Heywood

(1995) noted that biological diversity has been examined by as many as twenty different disciplines in the biological sciences, as well as others in the sociocultural dimension.

But scientists and society in general seem to have become enamored with the concept in the mid-1980s. Why did this occur? Increasing environmental appreciation and awareness at about this time appear to have been the catalyst that stimulated its growing interest. Wilson (1997) has noted that the first convincing estimates of the rate of tropical deforestation were made in the late 1970s and through the early 1980s. People began to ask whether they were in the midst of a global biodiversity crisis and how they would be able to tell if in fact they were. Scientists who had focused on specific details in biology became energized to realize that their research "mattered" in the larger scheme of life. A trend toward the holistic approach to investigation emerged, and interrelationships among disciplines became more typical. The "science" and study of biodiversity had come of age.

At the 1986 National Forum on BioDiversity, a group of scientists brought to the attention of the global society that things were not quite right in our world and that we had better do something about it fast (Wilson and Peter 1988). Since then the idea that biological diversity is threatened by human activities at an increasingly alarming rate has continued to be brought forward (Reid and Miller 1989; McNeely et al. 1990; Pineda et al. 1991).

The Convention on Biological Diversity, ratified at the United Nations Conference on Environment and Development in 1992, has been described as one of the most significant and far-reaching environmental treaties ever developed (Heywood 1995). About 150 countries that are now signatories to the convention have further popularized the biodiversity concept throughout the world. Through the Convention, governments affirm sovereign rights over the biological resources found within their countries while accepting responsibility for conserving biological diversity and using biological resources in a sustainable manner. The good news is that governments remain keenly interested in the Convention. Several countries—including Canada, Australia, Bulgaria, Chile, Germany, Norway, the Netherlands, the United Kingdom, Philippines, China, Vietnam, Indonesia, Brazil, and Costa Rica—have developed or are developing national biodiversity strategies or action plans.

For example, Canada has produced one of the first national strategy documents, which summarizes the history of biodiversity conservation, defines the basic elements of biological diversity, establishes a vision for the country in terms of future conservation, and provides a set of guidelines for meeting its biodiversity conservation goals (Environment Canada 1996). The strategy presents a vision for Canada of "a society that lives and develops as a part of nature, values the diversity of life, takes no more than can be replenished, and leaves to future generations a nurturing and dynamic world, rich in its biodiversity." Each Canadian is challenged to ensure that biological diversity is maintained and conserved in the years ahead. The strategy presents five basic goals and 146 strategic directions for governments, nongovernmental organizations, and individuals to contribute to the conservation of biological diversity. The need for cooperation among diverse interests is recognized as critical to achievement of the national goals.

As a follow-up to the 1992 Rio Conference and the Biodiversity Convention, the United Nations Environment Programme commissioned the Global Biodiversity Assessment in 1995 (Heywood 1995). The objective of the assessment was to

> provide an independent, critical, peer-reviewed, scientific analysis of the current issues, theories and views regarding the main global aspects of biodiversity. It assesses the current state of knowledge, identifies gaps in knowledge and critical scientific issues, and draws attention to those issues where scientists have reached a consensus of view and those where uncertainty has led to conflicting viewpoints and therefore a need for further research.

The assessment involved about 1,500 experts from around the world. The perspective is global and emphasizes general concepts and principles. Its goals are to provide assistance to countries addressing their specific biodiversity challenges and to provide a basis for decision making and further research in the field.

Additional initiatives in the biodiversity arena are continuing on a wide scale. The status of the field of biological diversity clearly is dynamic. The concept has a relatively recent history but is well founded in scientific thought. Its future development will no doubt continue to be influenced by scientific methodology as well as public opinion. It is our hope that its development pays heed to its past and strives to expand its future.

Ecosystem Management and Conservation Biology

The terms "ecosystem management" and "conservation biology" are not one and the same, although they have sometimes been considered as such. This unfortunate misconception has perhaps caused some confusion among natural resource managers who have been charged with maintaining biological diversity on public lands and encouraging private landowners to become partners in larger landscape-level management plans. Conservation of biological diversity is one objective of ecosystem management. Conservation biology focuses on this component specifically. Conservation biology, however, is a science or discipline that combines applied management principles, from fields such as forestry and wildlife and range management, with theories from the basic sciences to address problems of maintaining biological diversity. Meffe and Carroll (1994) described conservation biology as a discipline that applies principles from such theoretical fields as ecology, biogeography, population genetics, economics, and sociology to maintain biodiversity. Ecosystem management is an effort to conserve biological diversity while also meeting society's values, demands, and commercial needs. Thus, ecosystem management has three different objectives—social, economic, and ecological (biodiversity).

Individuals within disciplines such as conservation biology or wildlife ecology and management have struggled to implement ecosystem management approaches that will help maintain biological diversity while meeting social and economic demands of natural resources. Meeting these diverse objectives requires managers to look beyond a single species or single management objective approach for managing natural resources. Traditional natural resource management and planning has focused on understanding, quantifying, and managing the requirements of individual species or a single vegetation type, often by manipulating the structure and composition of forests, grasslands, or rangelands to achieve some human goal. These activities have largely been conducted by federal, state, and provincial natural resources agencies as well as by large private landowners. But these approaches to managing natural resources have been challenged as natural resource managers attempt to maintain biological diversity.

Maintenance of biodiversity requires managers to consider a broader perspective when developing management plans. Planning and conducting management activities to conserve biological diversity, however, should be viewed as complementing, not substituting for, species-level management or management for a single vegetation type. In this way, conservation of biological diversity can be viewed as a management goal that challenges us to make decisions over larger geographical areas. The larger geographical area may range from assemblages of stands to even a region of a state or part of a country, depending upon the spatial scale that needs to be addressed to maintain a spectrum of species or ecosystems.

Numerous definitions and descriptions of ecosystem management have been presented (for example, Grumbine 1994; Kaufmann et al. 1994; Haufler et al. 1996) elsewhere, but common components of many definitions include the need to maintain or restore biological diversity across all levels of biological organization, considerations of social needs and diverse public values within the landscapes of inter-est, and the economic outputs associated with natural resource management. Implementing a natural resource management process that is designed to meet the above goals could potentially aid in maintaining biodiversity and reduce the conflicts among the stakeholders involved in natural resource management issues.

Why Do We Need to Conserve Biological Diversity?

Management of natural resources has evolved with an improvement in our understanding of ecological relationships and a broadening of value for natural resources. As we enter the twenty-first century, biologists, ecologists, and natural resource managers will continue to manage landscapes to meet society's diverse needs but who will also need to maintain or enhance the integrity of our ecosystems. Some stakeholders who become involved with natural resource management issues want managers to continue to provide the consumptive and sustainable uses of natural resources for recreation and other direct human benefits. Others are primarily concerned with maintaining or "preserving" unique ecosystems for their existence values. Responsible resource management strives to achieve these and other types of stakeholder objectives. Moreover, to do so, the ecological importance and direct human benefits that biodiversity provides need to be communicated among stakeholders and natural resource professionals.

Ecological Importance of Biological Diversity

The ecological importance of biological diversity can be considered from a variety of different perspectives.

Ecosystem Composition, Structure, and Function

To maintain biological diversity at any level, it is essential to understand the compositional, structural, and functional characteristics of ecosystems. Understanding these characteristics allows managers to develop and prescribe management practices that minimize detrimental effects on biological systems while still allowing them to manage forests and grasslands to meet multiple objectives.

One key feature of all ecosystems is the interactions that occur between their biotic and abiotic components. These interactions provide functional features that characterize or distinguish one ecosystem from another. Each type of ecosystem also has a unique structure that allows energy and nutrients to flow among its various components.

Whenever a naturally or human-induced disturbance occurs within an ecosystem, its dynamic composition, structure, and functions change. In ecosystems such as those dependent on fire or fluctuating water conditions to alter plant communities, it is these disturbances that are essential for maintaining the diversity of other organisms associated with them. For example, how could the Kirtland's warbler (*Dendroica kirtlandii*) in Michigan be maintained without the harvesting or burning of mature jack pine (*Pinus banksiana*) stands? Optimal nesting and brood-rearing habitat for this species is provided by relatively young, large stands of jack pine (Line 1964; Anderson and Storer 1976). Harvesting mature trees allows for stands to regenerate or be replanted, and burning facilitates the opening and dissemination of seeds from cones.

Natural resource managers attempting to manipulate various habitat types to meet desired management objectives may prescribe management practices that emulate natural disturbances, such as controlled burns. By using these techniques, managers may meet their objectives while maintaining the compositional, structural, and functional characteristics of ecosystems. This, however, requires that they know how biological systems respond to ecological changes. In addition, they need to verify that forests and grasslands have not already been so altered by human-induced changes that emulated disturbances cannot be initiated without causing deleterious impacts. In some cases, natural disturbance regimes have been so altered by human actions that a return of those regimes without first preparing ecosystems for them results in, from a historical perspective, unprecedented, catastrophic events (Erickson and Toweill 1994).

Use of disturbances that alter the temporal patterns or the spatial scale at which disturbances normally occur or that infuse exotic species into various systems may be detrimental to the evolutionary development of some ecosystems. One example of how the infusion of exotic species and alteration of disturbance regimes can impact the structural and functional characteristics of biological systems has been discussed by Hanaburgh (1995) for remnant grasslands in Michigan. Hanaburgh (1995) concluded that it was likely that fire suppression has altered the vegetation structure of native grass-dominated sites as well as the wildlife community within those sites.

Species Interactions

Ecologists have not yet explored the majority of ecological relationships among living things. The absence of seemingly insignificant species within an ecosystem may have widespread implications for the existence of other species in an ecosystem. Some species may be more "essential" than others from a functional or ecological perspective and therefore are designated as "keystone" species (Terborgh 1986). Other species may be important as indicators of the integrity of the ecosystem in which they occur (Hunter 1990). For example, McNaughton (1975) discussed the role of migratory herbivores through the Serengeti in ungulate habitat quality and the structure and function of African grasslands. Understanding what roles species of herbivores play in maintaining habitat quality for other species and their functions within ecosystems may be valuable in land-use planning and natural resource policy intended to maintain biodiversity.

Genetic Variability

The genetic variability that exists in a population is critical to the population's ability to withstand stochastic changes in its environment. Populations that exhibit relatively higher levels of heterozygosity generally have a higher probability of surviving changes in habitat conditions and produce more offspring than populations that are more homozygous. For example, Packer et al. (1991) have noted that some 75 to 125 lions (*Panthera leo*) in Ngorongoro Crater in Tanzania are derived from an estimated 15 founder individuals. Because this isolated population was developed from a relatively low number of individuals, it has been characterized as having less heterozygosity than lions occurring in the Serengeti plains. Depending on the frequency and intensity of stochastic events that may influence the quality and availability of lion habitat in this area (intense tourism, for example) or an additional epizootic event, this population may be susceptible to extirpation because of its lack of heterozygosity.

Direct Human Benefits of Biological Diversity

All species have the potential to provide direct benefits to humans. Science is continually finding human uses for species of all types. The roles or uses that some species or habitat types play are just more noticeable to us than others. If existing or unknown species are to provide direct benefits to humans in the future, we therefore must look at new approaches to conserve them. Wilson (1992) has mentioned that useful products can only come from species we currently have, not extinct species. The ability of plants and plant communities to produce oxygen, energy sources, building supplies, foods, and medicines as well as a source of income for local people through ecotourism are just a few examples of how biological diversity can provide benefits to humans.

Medicines, Foods, Energy Demands, and Building Supplies

Ethnobotany is the study of how cultures use plant materials to address their daily needs. Examples of the extent to which plant resources are currently used for medicinal purposes are described by Farnsworth and Morris (1976). Since the 1950s, in the United States alone, approximately 25 percent of all prescriptions from pharmacies have contained active substances that are extracted from plants (Farnsworth and Morris 1976). On the basis of how useful many plant species have been in providing medicinal benefits in health care, we can assume that there are probably many species of plants, perhaps many still unidentified, that may provide medicinal benefits for diseases such as AIDS or various types of cancer.

The biological diversity of our planet also provides the foods we eat, the energy we use, and the homes we live in. However, the number of species providing for these needs is relatively limited. The Council on Environmental Quality (1981) estimated that approximately 80 percent of the world food supply is provided by fewer than two dozen species of plants and animals. Eckholm (1975) estimated that wood is used among all cultures to meet 90 percent of their energy demands. The demand for wood and wood products, a naturally produced renewable resource, is also essential for construction purposes. Reid and Miller (1989) estimated that the global timber trade in 1989 was in excess of $77 billion U.S. dollars.

Local Income

Where do people like to spend their leisure time? Oftentimes recreation occurs in areas associated with a relatively great diversity of species. Kenya, for example, largely depends on revenues brought into the country by tourists who come to see its diversity of birds, herbivores, and large cats. Describing the ecological impacts of tourism on wildlife and wildlife habitat in the Maasai Mara National Reserve, Muthee (1991) noted that the number of tourists in Kenya has increased from 35,000 in 1960 to 676,000 in 1988. Obviously, this is a development that can have both positive and negative impacts on the natural resources of the country. As tourists flock to many regions of the world to view unique ecosystems or species in their natural habitats, local people also are drawn to them to reap some of the financial benefits of having tourists visit their countries. In turn, these areas grow in population and become so crowded that they may function much like a large pen enclosing the animals inside them. Increased revenues mean there is funding to support the environmental values.

A challenge for maintaining the biological diversity associated with such unique areas as the Maasai Mara National Reserve will be to strike a balance among development and population growth, tourism pressure, and the need to maintain viable populations of wildlife species for viewing. If a balance is not developed, the quality of habitat in these areas will decline from tourism pressures and animal densities. Ultimately animal numbers will decline or the animals will not be able to disperse as they have traditionally without resulting in wildlife–human conflicts. And ultimately, as optimal wildlife viewing opportunities decline, so too will revenues from tourism.

Summary and Conclusions

Natural resource managers face a great challenge in trying to maintain and, in some cases, restore biological diversity. Maintaining biodiversity will aid in meeting diverse economic and recreational demands for natural resources while maintaining the integrity of ecosystems. Approaches to maintaining biodiversity may best be accomplished by implementing some form of ecosystem management.

Chapter 2

Strategies for Conserving Terrestrial Biological Diversity

Jonathan B. Haufler

One of the challenges facing natural resource managers today is how to maintain biological diversity while still meeting society's desires and needs for commodities, recreational opportunities, aesthetics, and all the other demands it places on resources. The recognized need to deal with these concerns in a unified manner has led to the application of ecosystem management. Ecosystem management has been accepted as the preferred approach for future land planning by most federal agencies, many state agencies, and a number of private landowners. The USDA Forest Service, for example, has focused much of its "reinvention" (USDA Forest Service 1994a) on needs for implementation of ecosystem management.

While definitions of "ecosystem management" vary, a common theme is that it must strive to meet ecological objectives as well as social and economic needs (Kaufmann et al. 1994). The maintenance and enhancement of biological diversity and ecosystem integrity are objectives of ecosystem management that are receiving considerable attention (Grumbine 1994). Various strategies have been proposed to achieve these objectives. The strategies are based on different assumptions and, in some cases, a different conceptual view of land management, and are currently being reviewed and debated.

The chapters in this book describe a number of strategies for conserving biodiversity. Additional strategies have been discussed elsewhere in the literature, and several of these are referenced here as well. This chapter presents an overview of the types of strategies available and discusses some of the assumptions of each.

Table 2.1. Selected Strategies for the Conservation of Biodiversity and a Comparison of Their Assumptions, Plans for Action, as Well as a Listing of Examples for Each Strategy

Strategy	Assumption	Actions	Examples
Bioreserve	Biodiversity declines are due to human activities	Restrict humans from selected areas to conserve biodiversity	Cooperrider et al. (Chapter 3); Noss and Cooperrider (1994)
Emphasis Area	Biodiversity depends on key areas of the land-scape maintained in appropriate conditions	Manage key areas to conserve biodiversity	Everett and Lehmkuhl (Chapter 6, 1996)
Coarse-Filter Habitat Diversity	Biodiversity depends on diverse habitat con-ditions	Maintain diverse habitat conditions to conserve biodiversity	Oliver (1992); Thomas (1979)
Historical Range of Variability	Biodiversity evolved with/adapted to con-ditions created by historical disturbance regimes	Maintain landscape within historical range of variability to conserve biodiver-sity	Aplet and Keeton (Chapter 5); Morgan et al. (1994)
Fine Filter	Maintaining all species will maintain biodiver-sity	Maintain appropriate conditions for all species/species guilds	Wall (Chapter 8)
Coarse Filter with Species Assessment	Biodiversity depends on adequate ecological representation of inherent ecosystems assessment of selected ecosystems	Identify adequate ecological represen-tation of coarse filter and check with an assessment of selected viability	Haufler et al. (Chapter 7; 1996)

Categories that have been designated for purposes of this discussion are bioreserve strategies, emphasis-area strategies, coarse-filter strategies based on either habitat diversity or historical ranges of variability, fine-filter strategies, and coarse-filter/fine-filter combinations (Table 2.1). Each strategy is presented in a manner designed to contrast its differences with the other strategies. One must keep in mind, however, that more moderate applications of each strategy are possible.

Terminology

Discussion of strategies for the conservation of biodiversity is frequently complicated by differences in terminology. Several key terms that are frequently confused are "coarse filter," "fine filter," "coarse scale," and "fine scale." As used here, "coarse filter" and "fine filter" refer to strategies for conserving biodiversity and are similar to the definitions of Marcot et al. (1994). Terms relating to habitat follow the recommendations of Hall et al. (1997). The scale terminology we employ follows the definitions recommended by Haufler et al. (in press).

Coarse filter. Strategies for setting biodiversity planning goals based on providing an appropriate mix of ecological communities across a planning landscape, rather than focusing on the needs of specific species.

Fine filter. Strategies for setting biodiversity planning goals based on the needs of individual species or guilds of species, thus providing for the needs of those species or guilds.

Resolution. The level of detail, such as pixel size or graininess, that is incorporated into the mapping of an area or in the collection of data.

Size or extent. The amount of area or length of time contained in a delineated landscape or time span or a measure of its breadth and width or duration.

Coarse scale. A level of resolution or grain size used in determining or mapping data based on larger units, such as a large pixel size or large grain.

Fine scale. A level of resolution or grain size used in mapping or measuring data based on units such as small grain size or small pixel size.

Broad scale. Analysis or management applied to a large area, for example, over a relatively large amount of acreage or over a long duration.

Small scale. Analysis or management applied to a small area, over a relatively small amount of acreage or over a short duration.

Habitat. The interacting physical and biological factors that allow an organism to successfully occupy an area.

Habitat type. Geographic area capable of supporting a certain vegetation type, usually the climax vegetation type for that particular area (Daubenmire 1968).

Vegetation type. A particular grouping of plants that typically occur together and have similar properties of composition and structure.

Stand. A specific identified area with relatively homogeneous structure and composition of vegetation.

Planning landscape. An ecologically delineated area of sufficient size that is able to contain viable populations of nearly all of the native species for the area, with the exception of a few megafauna or species with very large home-range requirements or consistently sparse population densities (Haufler et al. 1996).

All too often scale and approach become ambiguous and confused in the course of developing strategies for conserving biodiversity. A coarse-filter strategy may be based on data generated at either a coarse scale or a fine scale, but because of its assumptions it is most appropriately applied to a large-size or broad-scale area. Coarse-filter strategies, if based on fine-scale data, will generally involve large amounts of data and require significant data storage and management capabilities. For this reason, coarse-filter strategies are often based on coarse-scale data, though they can certainly operate using either scale of data.

Fine-filter strategies may use either coarse- or fine-scale data for any size area, although the habitat requirements of most species generally require at least moderately fine-scaled data to make meaningful assessments. Species-specific habitat descriptions based on coarse-scale data often are too generalized to ensure that the habitat requirements of a species will actually be met. Thus, fine-scale data is the usual basis for the application of fine-filter strategies.

Bioreserve Strategies

A number of authors have proposed the use of bioreserves as part of land management to maintain or enhance biodiversity (Harris 1984; McMahon 1993; Alverson et al. 1994; Noss 1994; Noss and Cooperrider 1994; Blockstein 1995; DellaSala et al. 1995, 1996; Cooperrider et al. Chapter 3). Some confusion and debate exist in the application of this approach because of the definition of "bioreserves," as exemplified by recent opinion articles by Noss (1996) and Everett and Lehmkuhl (1996). At one level, "bioreserve" can be defined as an area whose primary use is the maintenance or enhancement of biodiversity. In contrast, it may also be defined as an area protected from human influences, such as a wilderness or national park.

Some authors distinguish bioreserve strategies from ecosystem management. Alverson et al. (1994) promoted a bioreserve strategy while describing ecosystem management as "these hybrid management-public relations initiatives . . . predicated on the dominance of timber and other commodity products in planning efforts and therefore promote forest management tied to the active manipulation of vegetation." Other reports see bioreserves as part of an ecosystem management effort. For example, the forest plan for federal lands in the Pacific Northwest (FEMAT 1993) utilized bioreserves, termed "late-successional reserves," as a primary focus of a plan considered to be an ecosystem management approach.

The following discussion of bioreserves and that by Cooperrider et al. (Chapter 3) is based on the definition of "bioreserves" as designated protected areas that attempt to eliminate human influences from core areas to the greatest extent possible. The basic assumption of this view is that human influences are the cause of biodiversity declines, and therefore, to maintain biodiversity, areas need to be set aside to minimize human activities. McMahon (1993) described a program for "ecological reserves" in Maine by identifying timber harvest, oil and mineral exploration, camping and campfires, motorized and nonmotorized vehicle use, and construction of trails, roads, and service areas as incompatible activities. A goal of these bioreserve approaches is to designate areas of appropriate size and location to be set aside as reserves and to connect these set-aside areas with corridors of similar composition.

The bioreserve strategy as it is practiced in the United States assumes that biodiversity maintenance is best achieved by protecting core areas from human activities or disturbances. In many developing countries of the world, however, bioreserves offer a land management approach where the only alternatives may be either protection of an area or excessive human use because of an inability to adequately control human activities. Honadle (1993) discussed the need for bioreserves in many countries in the tropics. In the United States, public land management agencies can designate and regulate to a greater degree the level of human activity on most public lands without the need for formal bioreserve designations. Thus, the imperative for bioreserves in many countries may be quite different in the United States.

Scott et al. (1993; Chapter 4) discussed a program termed "Gap Analysis" that maps vegetation, wildlife species distributions based on models, and land management categories for a state. These maps are used to identify habitat conditions and biodiversity hot spots that are not "protected" in areas managed for natural values. The "gaps" in the coverage of "protected" areas are then targeted as priority areas for establishment of reserves for biodiversity.

One of the questions to be addressed in any bioreserve approach is how many bioreserves are needed, how large they need to be, and how they should be distributed to provide for the conservation of biodiversity. Other considerations include the types of uses compatible within the bioreserves, the ways natural disturbance regimes and successional change will be factored into bioreserve management, and methods to deal with past influences of human activities. Cooperrider et al. (Chapter 3) discuss the application of a bioreserve approach to the Klamath Ecoregion and provides some information concerning these questions.

Emphasis-Area Strategy

Everett et al. (1994) and Everett and Lehmkuhl (1996; Chapter 6) propose a land management strategy that they term an "emphasis-area approach." Under this approach, key areas of the landscape that support sensitive species or unique habitats are identified and designated as emphasis areas. The areas would then be managed to maintain their desired ecological values and would have flexible boundaries; moreover, the dynamic nature of their sustainable ecosystems would be recognized. The approach reduces what Everett et al. termed "administrative fragmentation" and avoids the setting aside of areas in bioreserves that often ignore the dynamics of the ecosystem. This strategy also allows management activities, including commodity production, as long as the activities are designed to maintain or enhance the desired ecological conditions within the emphasis area.

Emphasis areas may be thought of as a bioreserve in the broad definition of the term—an area whose use primarily reflects ecological objectives, but not the more commonly used definition described above. The assumption of the emphasis-area approach is that by identifying important areas for biodiversity across a landscape and managing those areas to maintain their biodiversity value, biodiversity objectives can be met.

Coarse-Filter Approaches

The goal of coarse-filter strategies is to provide for ecosystem integrity or biodiversity by maintaining a mix of ecological communities or ecosystems at an appropriate landscape scale (Kaufmann et al. 1994). These strategies assume that biodiversity and ecosystem integrity can be maintained if the correct mix of ecosystems or ecological communities is provided. Several approaches for coarse-filter strategies have been proposed. Two will be discussed here: management for habitat diversity and management within historical ranges of variability.

Strategies Based on Habitat Diversity

Oliver (1992, 1994) discussed a strategy for ecosystem management in the Pacific Northwest, which he termed "landscape ecosystem management." Under this strategy, active management of ecosystems would maintain a mix of stand structures across a landscape. By providing a balanced amount of four stand structures from early succession (stand initiation) to late succession (old growth) within each watershed, biodiversity would be maintained. The assumption of this strategy is that if at least a minimum amount (say 10 to 15 percent) of each stand structure is provided, then habitat diversity will be sufficient to support biodiversity.

Other authors have proposed additional designs for habitat diversity. Thomas (1979) discussed the maintenance of species richness through the maintenance of habitat diversity, which he described as the mix of successional stages occurring within plant communities (the potential natural vegetation) of a region. This is conceptually similar to Oliver's (1992) landscape ecosystem management approach, with the added component of identifying the ecological layer represented by plant communities. Similarly, Hoover and Wills (1984) associated wildlife species with seventeen forested plant series for Colorado. They combined these seventeen series into nine ecosystems, within which they identified five successional stages (grass-forb to old growth), and then assigned habitat-quality ratings to the different categories for selected indicator species. The assumption is that a mix of successional stages, characterized by the various types of plant communities or series, provides sufficient habitat diversity to support all biodiversity.

Strategies Based on Historical Ranges of Variability

Another coarse-filter strategy designed to maintain biodiversity and ecosystem integrity focuses on maintaining the mix of ecosystems across a landscape within the historical range of variability. Aplet and Keeton (Chapter 5) describe the concept of historical range of variability and how it can form the basis for identifying areas that need to be represented across a landscape in order to maintain biodiversity. The primary assumption of this approach is that the native species of a region adapted to and occurred within the historical range of ecosystem conditions, and that by maintaining ecosystems within this range, the needs of all species will be met (Risbrudt 1992; Morgan et al. 1994). Under this approach, the desired amount and description of each successional stage within an ecologically defined unit, such as a habitat type, can be defined based on an understanding of historical disturbance regimes that affected each ecological unit and the resulting stand conditions.

An example of this strategy has been applied in southern Idaho at the Deadwood landscape (USDA Forest Service 1994b). Management guidelines set the acceptable range of vegetation conditions to maintain each habitat type or fire grouping of stands within its historical range of conditions. There was recognition that many of the stand conditions were outside their historical range of variability. Desired stand conditions would be restored through the use of active management, such as thinning, prescribed fire, or a combination of both. The objectives of these management actions were to maintain ecosystem integrity and viability of selected species of management concern.

Fine-Filter Strategies

Fine-filter strategies attempt to maintain biological diversity by providing for the specific needs of individual species, guilds, or other groupings of species. Hunter (1990) has outlined various ways of selecting or grouping species to be targeted in fine-filter strategies. Arguments can be made that biodiversity cannot be maintained without considering the needs of the ultimate indicators of biodiversity, populations of individual species. The assumption of this strategy is that biodiversity can best be maintained by managing for the needs of all species by either considering species individually or by aggregating species into groupings such as guilds or life-forms. One reason for applying this strategy is that most legal requirements or mandates relating to ecosystem management or maintenance of biodiversity, such as the Endangered Species Act and the National Forest Management Act, focus on species needs.

Marcot et al. (1994) have provided several examples in which featured species provided a focal point for ecosystem management.

Certainly, the plan presented by FEMAT (1993), which focused on maintenance of the northern spotted owl (*Strix occidentalis caurina*) and other late-successional dependent species, was developed as a fine-filter strategy for ecosystem management. Huff et al. (1994) discussed a fine-filter approach based on providing for vertebrate habitat requirements in the Mount Hood area. This approach developed a relational database of 307 vertebrates associated with the Mount Hood National Forest, which were then grouped into thirty-three guilds based on similarities in environmental-use patterns. The idea was to analyze a landscape for suitable patches of habitat for each of the guilds and include them in planning processes. Wall (Chapter 8) describes a guild approach to landscape planning in northern Idaho.

Coarse-Filter/Fine-Filter Approaches

Coarse-filter and fine-filter approaches substantially differ in focus. Coarse-filter approaches identify ecological communities that can be incorporated into landscape planning. Depending upon the classification system used to identify ecological communities, successional trajectories, historical disturbance regimes, and ecological processes can all be factored into a coarse-filter approach. In addition, it is more feasible to plan for ecological communities than it is to plan for potentially hundreds or thousands of species. Fine-filter approaches track individual species or guilds. Hunter (1990) discussed how some species may "slip through the cracks" of a coarse-filter approach but may be protected in a fine-filter approach. Thus, fine-filter approaches address one of the basic components of biodiversity directly, that is, the species.

Combining these approaches can offer advantages. Haufler et al. (1996; Chapter 7) describes a coarse-filter approach with a species assessment. The assumptions of this approach are that to maintain biodiversity, sufficient amounts of all ecosystems occurring under historical disturbance regimes need to be provided. However, in contrast to the coarse-filter strategy based on historical range of variability described above, this amount can be less than the historical range of variability. As a check on this amount, assessment of the viability of selected species is conducted.

Hunter (1990) and Kaufmann et al. (1994) recognized the advantages of this type of combined approach, though they did not identify any specific framework for its application. The coarse-filter approach with a species assessment will require significant data to implement, in that the coarse-filter planning must incorporate sufficient fine-scale data to allow for an adequate assessment of the habitat for the selected species. It also relies upon the assumption that assessments of quantities and qualities of habitat can be linked to desired population responses by species targeted in the species assessment.

Categorization of Strategies for Biodiversity Conservation

While this categorizing of approaches may not address every proposed strategy for the conservation of biodiversity, it defines the philosophical approaches and assumptions behind the range of possibilities being discussed. These strategies are also not completely independent. Components of different strategies can be incorporated within each other. The combination of a coarse-filter approach with a species assessment discussed above already melds components of other strategies. A coarse-filter strategy may also target selected areas for their ecological significance and manage them for these values, thus incorporating much of the reasoning behind the emphasis-area strategy. Aplet and Keeton (Chapter 5) combine elements of a coarse-filter approach using historical range of variability as a target within a bioreserve approach.

The contrasts of these different approaches points out the range of possible frameworks that landscape-planning strategies may take. But it must be kept in mind that the strategy selected will greatly influence the outcome of management recommendations resulting from the strategy. Alternative planning prescriptions that are based on different levels or intensities of the same strategy will not adequately frame the true range of alternatives. For example, if a bioreserve strategy is selected, then "alternatives" that might be considered in management planning could be the varying numbers and sizes of bioreserves. In contrast, if a bioreserve strategy were identified as one alternative, a coarse filter based on historical range of variability as a second, and a coarse filter with a species assessment as a third, a very different set of management plans could be compared and contrasted. Alternative management actions planned across a landscape can be strongly influenced and even constrained by the framework for the selected strategy. Thus, consideration of the influence of a landscape strategy on the outcome of possible management alternatives is a critical step in a landscape-planning process.

All proposed landscape strategies represent theoretical approaches to landscape planning. Several are presently being implemented at various scales, as discussed in later chapters of this book, and many others are being considered. None has actually been fully implemented and tested for effectiveness. Therefore, all landscape strategies for biodiversity conservation must be viewed as working concepts and should be considered in an adaptive management approach (Holling 1978; Walters and Holling 1990; Nudds, Chapter 11).

Scale Considerations

All strategies for the conservation of biodiversity must take into account the appropriate scale at which they should be applied. Relevant scale considerations include identification of the appropriate extent of the planning landscape, the resolution of the mapping or data to be collected, and the duration of plans (Haufler et al., in press). Regardless of the strategy, the extent of the planning landscape must be relatively large if biodiversity is to be addressed in a meaningful fashion. Contributions toward the conservation of biodiversity, such as meeting the needs of one or more species or subpopulations of concern, can be addressed in landscapes of smaller extent, and these contributions can be aggregated to address broader goals of biodiversity conservation.

Additional factors relating to the extent of planning landscapes and the resolution of mapping or data collection that need to be considered in biodiversity conservation efforts include the types of historical and present disturbance regimes affecting ecological communities and the size and timing of their influences. These are further influenced by ecological boundaries defined by gradients in geology and climate. Ignoring the effects of these scale issues on the implementation of a conservation strategy will severely decrease the likelihood of success.

Haufler et al. (1996) recommended the use of planning landscapes based on the section or aggregates of subsections level of Ecomap (1993) and Maxwell et al. (1995). This level corresponds to Bailey's subregions (Bailey 1995, 1996). A planning landscape of this size was felt to be large enough to provide sufficient ranges of habitat conditions to allow viable populations of a large majority of the native species that historically occurred in the landscape. It would also be large enough to incorporate the range of historical disturbances and yet be delineated by ecological boundaries that influence the ecological communities and disturbance processes within the landscape. At this level, a coarse-filter approach should be able to provide for ecosystem integrity within the landscape and provide for viable populations of a large majority of species. A few additional species with very low population densities or very large home-range requirements may require aggregation of habitat units from multiple planning landscapes to assure viability, but the number of such species is small enough to be accounted for with a fine-filter assessment. A landscape of this size also provides an area that maintains sufficient ecological similarity in ecological communities for a coarse-filter strategy to be defined. Kernohan and Haufler (Chapter 15) further discuss the variance considerations with multiple planning objectives and scales.

An additional scale issue concerns time frames to be used in strategies for the conservation of biological diversity. Planning time frames need to be long enough to consider the disturbance regimes and successional processes affecting the ecological communities (Haufler et al., in press). At some point, however, lack of available data becomes a major limitation on analyses and extrapolations. For example, data on fire regimes that date back more than several hundred years are very difficult to obtain. At the other extreme, attempting to define biodiversity objectives relative to geological time spans causes "baseline" comparisons to lose context and adds little to current conservation efforts.

Science, Policy, and Risk in the Conservation of Biodiversity

Much of the focus of currently proposed or implemented efforts toward conservation of biodiversity is on the development and testing of scientific methodologies and tools for achievement of this objective. This is critically needed, as no established scientific foundations have been empirically tested. However, much of the debate and potential success of conservation initiatives depends upon the social and political acceptance of and agreement with any initiatives (Kernohan and Haufler, Chapter 15). As noted earlier, a number of different strategies for conserving biodiversity, with different underlying scientific foundations, assumptions, and outputs, have been proposed. Scientific discussion on the advantages or disadvantages of each of these will undoubtedly continue into the future. In the meantime, implementation and application of various strategies will be discussed within and among agencies and private organizations and in political settings. This raises the question of the appropriate role of science and scientists in the development of conservation strategies for biodiversity.

Clearly, scientists have a major role to play in developing and testing various strategies for conserving biodiversity. Scientists need to accept that they cannot eliminate all risks to biodiversity. The needs and demands of human society will continue to pose risks to biodiversity. Rather, scientists need to propose management strategies that are designed to conserve biodiversity and to accurately assess the level of risk to biodiversity associated with various levels or applications of those strategies. As part of this testing, scientists need to try to frame the level of variance, accuracy of information, and levels of risk associated with the planned outcomes of their proposed strategies. These assessments need to occur at all levels of the process, from assessment of the level of variance around the strategy itself to the variance in the data collected for implementation of the strategy to the resolution and accuracy of mapping ecological parameters. Only by objectively presenting accurate risk estimates can policymakers be provided with the best information for making administrative or legislative decisions. The level of risk to biodiversity can then be most effectively evaluated in relation to economic or social costs.

Terrestrial, Wetland, and Aquatic Communities

The emphasis in this book is on strategies for the conservation of biodiversity associated with terrestrial ecosystems. Some chapters discuss strategies that cross terrestrial, wetland, and aquatic ecosystems, such as bioreserve strategies (Cooperrider et al. Chapter 3). Other chapters present strategies that may be applied to terrestrial ecosystems but that could function equally well in wetland or aquatic ecosystems, as is discussed in the coarse-filter approach with a species assessment (Haufler et al. Chapter 7). Terrestrial systems have generally received more attention from the standpoint of biodiversity strategies. Bio-diversity conservation in wetland and aquatic ecosystems has more typically been approached from a prescriptive or protective perspective, rarely on a coarse-filter basis, although examples of additional aquatic and wetland strategies undoubtedly exist.

Some of the examples in this book focus on vertebrate species. Few specific examples of plants or invertebrates are included. Certainly many of the strategies described are designed to include all species in their conservation strategy, although specific examples often use vertebrate species. Thus, most of the strategies presented in this book should apply equally well to all species, including plants, fungi, invertebrates, and vertebrates.

Summary and Conclusions

Strategies for the conservation of biodiversity include bioreserve strategies, an emphasis-area strategy, coarse-filter strategies based on habitat diversity or historical ranges of variability, fine-filter strategies, and a coarse-filter/fine-filter combination, each based on a number of different assumptions and approaches. This book compares and contrasts many of these various strategies and their assumptions, advantages, and disadvantages. Because all of these strategies are currently experimental, it is critical that they be monitored and evaluated, most preferably in an adaptive management format. Scale considerations are critical in implementing any proposed conservation strategy. Also, it is important that the variability, accuracy, and level of risk of each strategy be carefully evaluated and articulated so that policies can be based on the best information. Finally, while this book emphasizes terrestrial ecosystems and vertebrate species, the strategies described are designed to address broader species groups and can and do work equally well for wetland and aquatic ecosystems.

$\mathcal{P} \mathcal{A} \mathcal{R} \mathcal{T}$ 2

Strategies for Conserving Biodiversity

Part 2 provides detailed descriptions of approaches used to conserve biological diversity in a broad array of management settings. Each chapter generally begins with a theoretical context of the approach, moves on to an exploration of specific techniques and assumptions, followed by an identification of complications and other issues, and ends with a discussion of advantages and disadvantages. Each chapter is set in a specific location, but we believe that application of the techniques to broader areas is possible and indeed necessary.

Chapter 3 describes the bioreserve strategy, a promising but largely untested concept for conserving biodiversity. In its most basic form, the strategy involves dividing the landscape into zones that range from total protection (no human activity) to what we can term "biodiversity sacrifice areas." Cooperrider et al. describe variations of the basic model and focus discussion on the Legacy Project in California. Advantages of the bioreserve approach are simplicity, applicability at any scale, familiarity of the zoning concept, and explicit provision of areas for human uses. Disadvantages include legal difficulties, inflexibility, need for large knowledge bases, political obstacles to new land-use restrictions, and a lack of experience with large-scale applications of the concept.

Chapter 4 discusses the rather formidable "diversity" of approaches and strategies that have been developed by natural resource managers to conserve biological diversity. While this variety might be considered admirable, Scott et al. note that it has made it difficult for managers to decide which strategies to choose for their own management situations or if a particular strategy is well suited to their specific setting. The authors present the argument for using all known elements of biodiversity—from genes to landscapes—in a hierarchical and iterative fashion. Application of this approach to any natural resource management situation is also described.

Chapter 5 presents the thesis that protecting species diversity requires us to maintain the landscape features that historically provided habitat diversity. Aplet and Keeton describe historical range of variability (HRV) in ecosystem characteristics as a model of habitat diversity that sustains species diversity. A coarse filter for biodiversity protection, the HRV is a conceptualization of the variability in ecosystem composition, structure, and function over multiple spatial and temporal scales. Managing habitat using HRV as a model means trying to maintain or restore fluctuations in measurable habitat factors, such as old-growth acreage, landscape patch size and juxtaposition, and stream flow and water quality, within the bounds of historical conditions. The authors believe that, in conjunction with a reserve-based strategy, HRV can help determine the character of an ecosystem that is "natural enough" on the managed lands outside reserves.

Chapter 6 discusses the emphasis-area approach, which makes the conservation of sensitive species and habitats a management priority within designated areas and encourages their conservation in adjacent lands. Everett and Lehmkuhl point out that sustainability of emphasis areas depends on management of disturbance effects at multiple scales, both smaller and larger than the core emphasis area. Management prescriptions for emphasis areas may require reduced or increased disturbance effects well beyond the core area. The boundary for emphasis-area consideration flexes with the scale of disturbance. Conversely, because emphasis areas are part of larger ecosystems, management activities that create or lead to long-term sustainability of the broader ecosystems may be encouraged within the emphasis area. The authors note that this approach reduces administrative fragmentation of the landscape by eliminating rigid preservation area boundaries. Emphasis areas become an integrated component of the larger landscape rather than isolated islands of preserved habitat.

Chapter 7 discusses how coarse-filter approaches to landscape management offer an operationally feasible method of planning at a landscape scale. Haufler et al. note that a concern with most coarse-filter approaches is that the needs of some species may be overlooked. Linking a coarse-filter approach with species assessments combines an operational approach with an assurance of meeting goals for regional biodiversity. A coarse-filter approach based on providing adequate ecological representation of ecosystems across a landscape as defined by an ecosystem diversity matrix is described. Species assessments based on habitat models using habitat variables in GIS coverages of the diversity matrix provide a check on meeting species needs. The authors conclude by suggesting that this strategy can be used to meet biodiversity goals in a flexible and sustainable manner.

Chapter 8 discusses the use of landscape management planning as a practical approach for managing an industrial land base to conserve biological diversity. Wall suggests that the necessity and demand for industrial land bases to respond to management of multiple resources have increased dramatically. The author describes a practical and implemented landscape management approach for a forest land base contained within a mixed-ownership operating area in north-central Idaho. This approach is bottom-up, built on forest stand information (4 to 60 hectares) aggregated to second- to fourth-order watersheds (1,000 to 2,000 hectares) aggregated to Landscape Management Units (8,000 to 24,000 hectares). The strategy includes watershed analysis, a coarse-filter/fine-filter habitat approach to maintenance of terrestrial species, and is rooted in an adaptive management process. Advantages and disadvantages are also discussed.

Chapter 9 describes the approach utilized by Ducks Unlimited in planning and delivering habitat programs on the basis of ecologically definable and contiguous landscapes. Neraasen and Nelson provide an analysis of a landscape conservation program called Prairie CARE, which is most advanced in the grassland and parkland regions of the Canadian Prairie Provinces. More than 100 landscapes in high-quality waterfowl production areas have been delineated by means of wetland and land-cover data. The authors describe the process of comparing waterfowl recruitment between a baseline landscape and one in which proposed changes in land use are expected to improve waterfowl productivity and increase biodiversity. Extrapolation of this approach to other landscapes is also discussed.

Finally, Chapter 10 discusses the need for a "big picture" view of biological diversity conservation. Samson and Knopf point out that, in many management situations, this perspective has not been recognized or is missing because basic principles have been applied inappropriately. Our ecological understanding has shifted from one of interactions within small areas and habitats to one based on a variety of broad-scale ecological and evolutionary processes, historical events, and biogeographical circumstances. Their approach recognizes that management of diversity is of little value without a description of the ecological processes that determine diversity. The authors draw on examples from the Great Plains of North America and other major biogeographical regions to illustrate the approach and the need to emphasize ecological processes.

Chapter 3

The Bioreserve Strategy for Conserving Biodiversity

Allen Y. Cooperrider,
Steven Day, and
Curtice Jacoby

The bioreserve strategy is a promising but largely untested approach to conserving biodiversity. The strategy involves zoning regional landscapes into areas that range from total protection (minimal human activity) to areas of intensive human use. Zoning, in this context, does not necessarily refer to a formal regulatory designation, but rather to a societal agreement to limit certain human activities and uses on certain lands. This agreement may be expressed and played out in a variety of ways, ranging from formal designation as reserves or parks to conservation easements or landowner agreements.

In this chapter we:

- review the development of the bioreserve strategy
- briefly describe the elements of the strategy and the context in which it should be applied
- discuss some strengths and weaknesses of the approach
- describe a case history of an initiative to apply the strategy to a portion of an ecoregion in Northern California.

Development of the Bioreserve Strategy

The concept of bioreserves has been around in one form or another for hundreds, if not thousands, of years. Some indigenous cultures recognized areas in which human activities such as hunting were forbidden. More recently the movement to create national parks, national wildlife refuges, wilderness, and natural areas is derived, at least in part, from the recognition that areas in which human activities are constrained are needed to conserve plant and animal life.

Why Bioreserves Are Necessary

Bioreserves serve at least three important biological functions: (1) the preservation of large and functioning ecosystems; (2) the preservation of biodiversity; and (3) the protection of specific species or groups of species (Meffe and Carroll 1994). Given a definition of "biodiversity" such as "the variety of life and its processes" (Keystone Center 1991), all three functions can be considered to have the purpose of "preservation of biodiversity." In fact, many conservation-oriented individuals and organizations believe that protected areas are among the most valuable management tools for preserving genetic material, species, and habitats and for maintaining various ecological processes (Christensen et al. 1996:689–691; McNeely et al. 1990:56–62; Primack 1993:326; Reid and Miller 1989:67).

Bioreserves serve another important function: they are benchmark areas against which we can judge how well we are maintaining biodiversity on the managed landscape (Christensen et al. 1996). They are essentially control areas for management experiments. This value was recognized by Aldo Leopold (1941, 1949), who pointed out that wilderness provides a "base datum of normality" for a "science of land health." Thus, bioreserves serve not only a preservation and conservation function by themselves but also an informational function for judging the effects of management on the larger human-managed landscape. As the Ecological Society of America has stated, "protection of natural areas in reserves is an essential component of an overall ecosystem management plan" (Christensen et al. 1996).

Problems with Existing Bioreserve Systems

Nevertheless, this approach of setting aside areas as parks, refuges, and so on has proved to be inadequate for at least three primary reasons (Noss and Cooperrider 1994). First, most such areas have been selected for purposes other than protection of biodiversity, or with limited consideration of this as a purpose. For example, most wilderness areas have been set aside based upon their fortuitous lack of roads together with their high value for primitive recreation, resulting in a plethora of high-altitude reserves but with few located in lowland areas. Similarly, many national parks and monuments in the United States have been selected because of their spectacular geological features rather than because they represented or conserved biotic communities.

Second, many species and plant communities are not represented or are underrepresented in existing reserve systems (Crumpacker et al. 1988). Given the manner in which reserves have been designated, this is not surprising.

Finally, most reserves are too small to contain fully functional biotic communities—that is, they are too small to contain natural disturbance regimes such as fires and too small to include year-round home ranges of larger, more mobile species such as migratory birds or migratory ungulates. For example, the oldest and largest national park in the conterminous United States, Yellowstone, is not large enough to support viable populations of many species (Clark and Zaunbrecher 1987). Furthermore, there is scientific evidence that national parks are losing species and that such loss is correlated with their size, the smaller ones losing more species than larger ones (Newmark 1985).

In summary, the reserve system that we have in the United States today was developed in an ad hoc, piecemeal manner for a variety of purposes, rather than with systematic consideration of the need to protect biodiversity. As a result, we have a system that is inadequate in terms of size and number of reserves.

The Bioreserve Strategy

To counter these limitations, conservation biologists are developing and testing ways to improve upon the past approach. Three major modifications have been suggested. First, and most central, is the systematic design of regional reserve systems. This process includes both designation of new reserves and improved design of existing reserves. A key element of the process is a systematic effort to ensure that all native species of plants and animals and all ecosystem types (biotic communities) are included in a reserve system. A second improvement is the "buffering" of core reserves (those areas where human activity is most constrained) from areas of increasing human activity and impacts. This strategy increases the effective protection of the core reserves. Finally, the utility of reserves can be made more effective by connecting core reserves with corridors or other forms of connectivity that allow some movement of plants and animals among core areas.

These concepts of systematic design of reserve systems, buffering of core reserves, and connectivity are the central principles of the "bioreserve strategy," as proposed by Noss (1983), Noss and Harris (1986), Noss (1992), Noss and Cooperrider (1994), and others. The application of these principles to achieve biodiversity goals is described briefly here and in more detail in Noss and Cooperrider (1994).

Goals and Objectives

The goal of a bioreserve strategy is to maintain the biodiversity of a region in perpetuity. Four fundamental objectives follow from this goal:

1. To represent, in a system of protected areas, all native ecosystem types and seral stages across their natural range of variation.

2. To maintain viable populations of all native species in natural patterns of abundance and distribution.

3. To maintain ecological and evolutionary processes, such as disturbance regimes, hydrological processes, nutrient cycles, and biotic interactions.

4. To manage landscapes and communities to be responsive to short-term and long-term environmental change and to maintain the evolutionary potential of the biota. (Noss and Cooperrider 1994).

Proponents of a bioreserve strategy recognize that these objectives can be achieved only with support of the local people over the long term. Therefore, it is important to emphasize that the bioreserve strategy also offers a systematic approach to providing for the social, economic, and spiritual needs of the local people while achieving biodiversity objectives.

Design Components

Fundamental to the bioreserve strategy is the concept of zoning the regional landscape to distinguish areas into areas with varying intensity of human activities. These consist of four major components: core reserves, buffers, zones of connectivity, and matrix. These categories are described briefly below and in more detail in Noss and Cooperrider (1994).

Core reserves. Core reserves are the backbone of a regional reserve system. They are areas in which the overriding goal of management or stewardship is preservation of native biodiversity and ecological integrity. They may consist of all or portions of national parks, wilderness areas, research natural areas, state parks and preserves, national wildlife refuges, and other areas in which conservation of biodiversity is given a priority and human demands upon the landscape are limited or discouraged. Ideally, the reserves of a region should collectively contain all native ecosystem types and seral stages as well as all native species. The question of what is a "native" ecosystem type or species is a subject of continuing debate; however, good operational definitions of the terms "native" and "natural" are available (see, for example, Maser 1990). For most situations, a practical definition of a native species is one that arrived in a place on its own (without aid of humans).

Buffers. Buffer zones surrounding the core reserves are designated to increase the effectiveness of core reserves. These consist of zones in which increasing amounts of human activity and disturbance are allowed (Figure 3.1). For example, the first buffer zone would allow only light (nonmotorized) recreation, and the second zone would allow for motorized travel on roads but no logging or mining.

Zones of connectivity. A second method of increasing the effectiveness of core reserves is to ensure that there is connectivity between them. "Connectivity" here refers to the state of being functionally connected by movement of organisms, material, or energy. Zones of connectivity, if designed properly, allow for the movement of plants and animals and their genes from one core reserve to another (Shafer 1990). There are many forms of connectivity (see Noss and Cooperrider 1994:150–156). Corridors, which are typically long and linear, provide one form of connectivity but not the only one. Different forms of connectivity and their relative values are discussed in detail in standard texts on conservation biology (see, for example, Meffe and Carroll 1994:284–288).

Matrix. Finally, the matrix contains all the rest of the land that does not fall into one of the above categories. In this area, human uses and

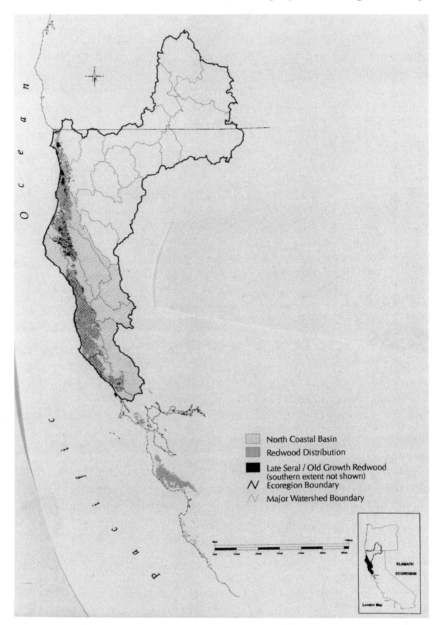

Figure 3.1. Map of Klamath Ecoregion showing the North Coastal Basin and distribution of coast redwood (*Sequoia sempervirens*).

demands upon the landscape receive priority in the traditional manner by which land is managed. The matrix may consist of agricultural lands, timberlands, pasture land, and urban and suburban development.

Strengths and Weaknesses of the Bioreserve Strategy

The bioreserve strategy is for the most part an untested strategy: we do not know how effective such an approach will be in conserving biodiversity over the long term (a hundred years or more). Nor do we know how effective such an approach can be in resolving conflicting human demands upon the regional landscape. Thus, the advantages and disadvantages of the bioreserve strategy described below represent a preliminary assessment of its practicality.

Strengths

An important characteristic of the approach is the simplicity of the concept. The basic concepts of core reserves, buffers, and zones of connectivity are easily understood, even though the methodologies and details of design may be quite complex. Furthermore, this landscape zoning concept can be applied at various scales ranging from the watershed to the ecoregion.

A related strength is that people are familiar with and, to varying degrees, accepting of the practice of landscape zoning. In spite of the rhetoric of private property rights advocates, land ownership in this country has never conferred an absolute right to do anything one wants on a property. This is true under both U.S. law and its predecessor, English common law.

Finally, by zoning the landscape into zones of restricted human activities and areas of intensive human activity, the bioreserve approach explicitly provides for areas of concentrated human activity. Much opposition to conservation efforts appears to come from the fear that more and more human activities are being restricted. By developing a "whole landscape" strategy, provision for human needs is made at the same time that some human activities are being restricted.

Weaknesses

The strategy is not without its pitfalls and limitations. Landscape-level strategies that even hint at restricting human activity on private lands raises fear of increased government regulation and "takings." An additional problem is the inflexibility of landscape zoning with set boundaries. Selecting, building public support for, and designating a core reserve can be a contentious and time-consuming process. If future information suggests that the reserve should have been in the next watershed, it may be difficult to change the design and make the necessary adjustments.

A major difficulty with a bioreserve approach is that it requires a large knowledge base, including basic information on

- the abundance, distribution, habitat requirements, and movement patterns of ecologically important (keystone, umbrella, threatened, and endangered) species. This information is needed to ensure that all species are included in the reserve system during the process of designing it.

- the distribution, abundance, and condition of all ecosystem types and their seral stages throughout the ecoregion and an estimate of their "natural" landscape pattern. As with the above, this information is needed to ensure that all ecosystem types are included in the reserve system when it is being designed.

- the region's natural disturbance regimes—the temporal and spatial scales at which they occur, how they have been modified by humans, and how they can be mimicked in a seminatural setting. This information is needed in order to determine size and placement of reserves and whether reserves can encompass disturbance regimes. Because most reserves are not large enough to contain disturbance regimes such as fire, it is important to know how to manage adjacent lands in a way that can mimic the natural disturbance regime.

- the area and level of protection of existing (de facto and de jure) reserves. These areas provide the building blocks for regional reserve networks.

This is a large amount of information to acquire, synthesize, and digest, but it represents only the most basic needs for biological information. Ideally, much more information on biology and natural history, land use, and land-use impacts would be available and could be used.

Finally, as alluded to earlier, the bioreserve strategy is a largely untested approach at the scale of the large regional ecosystem or ecoregion. We not only have little experience in applying these concepts at this spatial scale, but the relevant time frame for testing the success of such a system at the ecoregion level is decades or centuries. Thus, it will be centuries before we truly know how successful the strategy has been. However, this caveat applies equally well to all other regional biodiversity conservation strategies. We may learn pretty quickly if they are not working (by observing species extirpations and extinctions and other obvious signs), but proving success will require patience.

Applying the Bioreserve Strategy–The North Coastal Basin of the Klamath Ecoregion

We describe here a case history of an ongoing effort to apply the bioreserve strategy to a portion of the Klamath Ecoregion in northern California and south-central Oregon. This represents one attempt to apply such a strategy to a larger regional landscape. It differs from other efforts in other regions or even parallel efforts to apply a bioreserve strategy within the same ecoregion. We present it here as an example of an application of the bioreserve strategy that is tailored to the region and its biological and sociohistorical context.

Background–The Klamath Ecoregion and North Coastal Basin

The Klamath/Central Pacific Coast Ecoregion (Klamath Ecoregion) is one of fifty-two ecoregions defined by the U.S. Fish and Wildlife Service (U.S. Fish and Wildlife Service 1994). It is located in northwestern California and south-central Oregon and consists of all the watersheds or hydrobasins that drain into the Pacific Ocean from San Francisco Bay north to the Smith River (Figure 3.1). Thus, its boundaries are defined in terms of watersheds rather than some other criteria such as geology or vegetation type.

Any ecoregional delineation is bound to be arbitrary. Some interactions occur among ecoregions, and thus any delimited area is not going to encompass all species or ecological processes. However, the Klamath Ecoregion boundaries are biologically meaningful in many ways. They encompass most of the range of coast redwoods. Comparison of a map of potential natural vegetation (Kuchler 1964) with that of the ecoregion indicates that it contains virtually all of the potential natural sites for one vegetation type (pine-cypress forest) and most of the sites for four others (redwood forest, California mixed-evergreen forest, montane chaparral, and fescue-oatgrass) (U.S. Fish and Wildlife Service 1997). Together, these four types make up more than 50 percent of the ecoregion.

Within the Klamath Ecoregion are three subregions—the Coast Ranges, the Klamath Mountains, and the Modoc Plateau—which have relatively distinct geologic origin, climatic pattern, and vegetation. The North Coastal Basin is a region defined by the California Water Quality Control Board and is generally synonymous with the Coast Ranges geologic province. In terms of watersheds, it is defined as all the hydrobasins draining into the Pacific Ocean south of the Klamath River and north of San Francisco Bay (Figure 3.1). High rates of uplift, high rainfall, and unstable rock types in this region produce exceptionally high natural sediment yields (Mount 1995:241). Logging and grazing within these watersheds have exacerbated the existing conditions, producing some of the highest sediment yields measured in the United States (Mount 1995:241).

The area's best-known natural and scenic features are the redwood forests and the rugged coastline. The most notorious residents of the region are the northern spotted owl (*Strix occidentalis*) and marbled murrelet (*Brachyramphus marmoratum*) that have received so much notoriety in recent years. However, the region contains many lesser-known features and species, including unique coastal prairies and numerous species of endemic plants. The rivers once supported six species of anadromous salmonids (Moyle 1994) as well as numerous lesser-known species of fish and other vertebrates.

The redwood "rainforest" that most characterizes this region is unique in that it resides on the edge of the Pacific Ocean in a region that has a basic Mediterranean climate—that is, most of the rain comes in the winter months. This forest is a relict of more widespread Arcto-Tertiary forests that once covered much of the West (Baker 1984). For most of the region, the months of June, July, and August (at a minimum) are virtually without rainfall. Being adjacent to the ocean, however, the region is regularly covered by fog during the summer months. Much of the effective precipitation in this redwood/fog belt comes from the phenomenon of "fog drip"—the ability of the redwood trees to capture fog from the air and transport it to the ground, where it is utilized by many other life-forms (Becking 1982; Dawson 1996). Dawson (1996) has written:

> [T]he hydrology and therefore the ecology of redwood forests is intimately linked to the presence of the tree canopies themselves and the role they play in stripping fog, ameliorating the forest microclimate, and increasing the total annual income from water. Secondly, and perhaps most importantly, from a management perspective, is the fact that loss of redwood trees due to natural disasters . . . or from logging . . . which convert the forests to open habitats will dramatically alter the hydrological and ecological balance of these forests. . . . Loss of the canopy tree (*S. sempervirens*), therefore, would mean not only the loss of the biomass, nutrients within the biomass, and the soils (by postdisturbance erosion), but also a fundamental conversion of a once moist, cool, forested ecosystem into a more drought prone, and warmer ecosystem. (Dawson 1996:92)

In summary, intensive forestry, as practiced in the region, has not only the known effects such as erosion and nutrient loss, but can also fundamentally decrease the capability of the ecosystem to capture moisture from fog during the hot, dry months of summer. Overall, considering the unstable soils and relictual forests with unique flora and fauna, the North Coastal Basin is a region that is highly sensitive to land-use impacts.

Unlike other portions of the ecoregion, which contain more than 80 percent public (mostly federal) land, the North Coastal Basin consists of approximately 90 percent privately owned lands, the major owners of which are corporate timber companies. Ever since 1917, when a highway was punched north from San Francisco through the redwood region to the Oregon border, conservationists have been working to protect old-growth redwood forests (Angle-Franzini 1996; Leydet 1969). In spite of their efforts, only 10 percent of the land area within the natural range of coast redwood remained as old-growth redwood forest in 1986 (see Figure 3.1). Of this 10 percent, approximately 37,600 hectares (approximately 46 percent) are on private lands. This private-land acreage contained approximately 3.5

billion board feet of old-growth redwood timber, of which 1.5 billion board feet were removed between 1986 and 1996, leaving approximately 2 billion board feet on private lands (Fox 1996).

In recent years, concern has shifted to impacts on second-growth forests. California's commercial forests are logged three times faster than the national average (Jensen et al. 1990), and on privately owned industry lands, harvest each year exceeds growth by 22 percent (Jensen et al. 1990). Private land sales and corporate takeovers are emerging as a threat to sustainable management because sustainable management results in an undervaluation of timber holdings at current discount rates compared with rapid extraction and sale of the trees (Jensen et al. 1990:88). Thus, businesses tending toward sustainable management are ripe for takeover by companies intending to extract resources quickly (Jensen et al. 1990:88). As a result, many corporate timber companies lack incentives for sustainable management of timber resources, much less comprehensive management to conserve biodiversity.

California has a Forest Practices Act (Z'Berg-Nejedly Act of 1974) that theoretically regulates forest practices on private lands to ensure that they produce "maximum sustained yield of high quality forest products." In spite of this regulation, timber inventories on industrial forestlands on the North Coast have steadily declined since then and are projected to continue to decline until at least 2020 (California Department of Forestry and Fire Protection 1988:C27). Comprehensive data are difficult to obtain, but available evidence suggests that from a purely economic point of view these forests are not nearly as productive as their potential. For example, in Mendocino County, in the center of this region, board-foot production from commercial timberland could have been two to three times as high as at present if the forests had been managed for long-term rather than short-term profit and if harvest had not consistently exceeded growth (Burkhardt 1994).

From an ecological perspective, such intensive logging and road building not only reduce biomass but also harm terrestrial and aquatic habitats through (1) loss of old-growth habitat, (2) fragmentation of remaining mature stands, (3) displacement of species dependent upon mature trees or dead wood, (4) alteration of tree stand composition, (5) long-term loss of soil fertility, (6) removal of riparian vegetation, (7) soil erosion and disturbance, (8) streambank erosion, (9) altered streamflow regimes, (10) loss of stream shade, and

(11) increased sediment load in streams (Jensen et al. 1990:86). Unfortunately, the state lacks the information and biological review capacity to consistently assess the effects of harvest activities on wildlife species and their habitats (California Board of Forestry 1990; Jones and Stokes Associates, Inc., 1989; LSA Associates, Inc., 1989). The net result of all of these factors and others is that the North Coastal Basin ecosystem is severely degraded. The following are key problems and evidence of such degradation:

- Redwood forests have been cut to the point where only a small percentage of the original area remains old growth, and that continues to be cut (Fox 1996).

- All of the anadromous fisheries of the region are in decline (Moyle 1994).

- Virtually all of the rivers of the region have been declared "impaired," primarily by sediment, under Section 305b of the Federal Clean Water Act (California Water Quality Board 1995).

- Coho salmon have been declared "threatened" by the National Marine Fisheries Service throughout the region (National Marine Fisheries Service 1996, 1997a), and steelhead (*Oncorhynchus mykiss*) are also listed as threatened in part of the region and are being considered for listing in other parts (National Marine Fisheries Service 1997b).

- Numerous other species of plants and animals of the region are formally listed as either threatened or endangered or are in some way "at risk" although not yet formally categorized as such (U.S. Fish and Wildlife Service 1997).

Although there has been severe disruption of many of the forest and riverine ecosystems in this region, in some ways the region has more potential for ecological recovery than many other parts of California. This is primarily because of the lower human population density. The four counties that make up the total ecoregion have a population of 600,000, of which more than 60 percent is concentrated in the southernmost county (Sonoma), which sits next to the greater San Francisco Bay area.

The unique challenges of developing a bioreserve strategy in this region thus revolve around the following regional characteristics:

- Limited acreage of public lands or existing reserves and high percentage of private lands
- High percentage of lands owned by corporations with little incentive for long-term stewardship and ineffective regulatory enforcement
- Severely degraded forest and riverine ecosystems
- Ecosystems relatively sensitive to human disturbance because of relictual vegetation and inherently unstable soils combined with Mediterranean climate.

Obstacles to Biodiversity Conservation

The obstacles to conserving biodiversity in such a region are numerous. From the point of view of those concerned with conserving regional biodiversity, they include (1) strong and increasing demand for commodities, (2) degraded and fragmented habitat, (3) ineffective regulation of forest practices and other potentially degrading activities, and (4) fragmented regulatory authorities combined with a large cohort of complacent or uninformed citizenry. Problems of implementing a bioreserve strategy in such a region are numerous and are shared by many regions (Trombulak et al. 1996).

On the other hand, some attributes of the region differ considerably from those of many other areas where regional biodiversity conservation strategies are being developed. Most central is the high percentage of private land and the high percentage of corporate ownership. Many of the ongoing bioreserve strategies such as those described by Pace (1991), Noss (1993), Vance-Borland et al. (1996), and others are for regions with a high percentage of federal or other public lands. In these cases, the strategies can rely heavily upon linking up existing reserves (in the form of national parks, wilderness areas) with other federal lands currently used for multiple purposes.

Overcoming Obstacles

Given the nature of the region, many citizen conservationists and biologists both within and outside government have realized that conserving biodiversity within the ecoregion will require a long-term, systematic, and cooperative effort among citizens, government, academia, and nongovernmental organizations (NGOs). Furthermore, many have realized that in a region with such fragmented land ownership, it is unlikely that a single government agency will take the lead in the preservation effort. Thus, a need was recognized for an NGO that could provide a leadership role in developing and implementing a bioreserve strategy for the region.

In response to this need, a nonprofit 501(c)3 organization, Legacy—The Landscape Connection, was formed in 1993. Its mission is to provide information for protection and restoration efforts in the Klamath Ecoregion and promote conservation of native biodiversity through integration of local knowledge and science. A keystone project of the group is the development and implementation of a bioreserve strategy for the North Coastal Basin, which is viewed as a long-term endeavor. As an NGO, Legacy can play several key roles in implementing a bioreserve strategy:

- *Visionary.* Legacy is developing and disseminating a vision of a sustainable regional ecosystem and a means to move toward that goal.

- *Catalyst.* By collecting, analyzing, synthesizing, and disseminating information on biodiversity problems and ecosystem needs, Legacy is serving as a catalyst to stimulate government agencies to fulfill their statutory environmental responsibilities.

- *Partner.* Legacy is developing working relationships with natural resource and regulatory agencies, with other nongovernment organizations, and with watershed groups to pursue mutual goals.

- *Archivist.* Legacy is archiving critical biodiversity information not being kept elsewhere because of fragmented ownership and jurisdictional problems.

- *Educator.* Legacy is developing educational programs about important concepts of conservation biology and important ecological processes of the ecoregion.

Development and Implementation of the Strategy

Johns and Soulé (1996) outlined steps for implementing a bioreserve or wildlands reserve program. Legacy's program emphasizes five aspects that may differ from programs in other regions: (1) "soft reserve design," (2) alliance with watershed groups, (3) use of geographic information systems both for data analysis and for education, (4) ever-expanding partnerships, and (5) education.

Soft Reserve Design

The soft reserve design process begins with a draft with "soft" lines outlining general areas where new reserves are needed and where buffers and zones of connectivity need to be established in the region. The lines are then modified and finalized based on knowledge of people living and working in the watersheds. The willingness of landowners to exchange their land holdings for another location, to shift their land-use goals, or to conform with the regional strategy as needed is explored along with the biodiversity inventory. This procedure allows Legacy to introduce the concept of a regional reserve system in a manner that is less threatening to local residents and landowners.

The Watershed Connection

Communication links and alliances are being forged with local watershed and basin groups working to produce local biodiversity conservation plans. This interaction is a two-way process with more than twenty groups working on such plans in the region. Legacy provides information on regional issues and problems and how local watershed plans can be congruent with regional reserve design. Legacy is prepared to assist local decision making in the placement of reserves, buffers, and corridors and in design of citizen sampling and monitoring programs. This allows watershed groups to provide Legacy with more detailed information on critical areas for biodiversity conservation as well as de facto reserves—that is, areas in which biodiversity is being protected through conservation easements, other legal instruments, or simply through landowner stewardship. Thus, the ultimate implementation of a bioreserve system will come from local watershed plans as building blocks for an integrated regional plan.

GIS as Medium and Message

Legacy is relying heavily upon geographic information system (GIS) technology to efficiently manipulate place-based data and assist in analysis of biological and spatial relationships. However, maps displaying regional and watershed-level conditions are also an extremely valuable communication tool (Cooperrider et al., in press). GIS maps produced by Legacy are some of the most effective tools for communicating with citizens, other organizations, and government agencies. Legacy is prepared to provide watershed groups with similar GIS capability by either assisting them in doing local GIS work or advising them in setting up GIS systems.

Other Partnerships

Legacy is forming alliances with a variety of other organizations and agencies that share all or part of its biodiversity conservation goals. These range from state and federal agencies with broad, sometimes conflicting, mandates, such as the California Department of Fish and Game, to nongovernmental organizations with relatively specific missions, such as the California Native Plant Society.

The U.S. Fish and Wildlife Service is playing a key role in developing a holistic strategy for restoration of the Klamath Ecoregion. This strategy describes the various ecological issues that need to be resolved for effective restoration of the lands, waters, and biota of the ecoregion. It recognizes that ecosystem restoration cannot be accomplished by one agency or organization working in isolation. Rather, it proposes that biodiversity restoration must be accomplished by the diversity of agencies, NGOs, and private citizens working in tandem or in a coordinated fashion. Legacy's role as an NGO pursuing a regional biodiversity conservation plan is completely congruent with the service's ecoregion restoration strategy.

The U.S. Fish and Wildlife Service has also provided a critical support service for conservation efforts in the region by setting up a virtual GIS lab at Humboldt State University (Carlson et al. 1994). This lab is developing seamless GIS layers of land ownership, soils, vegetation, plant and animal distributions, and additional basic map data for the entire ecoregion. These GIS layers are being made available to citizens, NGOs, and other agencies throughout the region. This service encourages the resolution of ecological issues because all parties have access to the same information and it greatly facilitates many of the tasks that Legacy is pursuing.

Ultimately, biodiversity restoration in this region will require participation of the corporate timber companies. Unfortunately, at present, there seems to be little common ground or incentive for conservation groups and companies to work together. There are some signs that this may be changing, particularly with some of the smaller timber companies. The listing of coho salmon as threatened and the imminent listing of a number of other species of fish and wildlife impart a stronger need for all parties to work together to achieve both common and individual goals. One approach being tested in this region now is to develop mutually acceptable conservation plans for watersheds rather than at the regional or companywide level.

Education

The importance of education and information in achieving any of these goals is paramount. Legacy is pursuing an active effort to develop and disseminate educational modules throughout the ecoregion. Current efforts include modules on (1) principles of conservation biology, (2) reserve design strategy, (3) the role of large carnivores in the ecosystem, (4) the hydrologic cycle and the fog drip connection, (5) salmon as ecological, economic, and spiritual keystone species, (6) watershed protection and the citizen's role in it, and (7) the role of hardwoods in the forest ecosystem.

Summary and Conclusions

The key to understanding the potential of the bioreserve strategy is to recognize that it is a long-term strategy, not a quick fix to immediate problems. Notwithstanding its serious limitations, it remains one of the most promising approaches to conserving biodiversity. The reason for this assertion is philosophical. This optimism is based on the belief that a strategy that tries to rely upon natural ecosystem processes and components—developed through thousands of years of coevolution of plants, animals, and landscapes—is more likely to be successful in the long term than one that assumes that humans can do a better job of "managing" nature.

Bioreserves are not intended to serve as a stand-alone strategy for conserving biodiversity. Even though the bioreserve strategy allocates matrix lands as areas where human development and activities can be emphasized, these lands must also be treated with a certain amount of care. If, for example, agricultural practices within the matrix result in accelerated soil erosion, it will be detrimental to the entire regional ecosystem. Similarly, a bioreserve strategy must be complemented by many of the existing programs, such as those for protecting clean water and clean air.

The case history of the North Coastal Basin and the program of the nongovernmental organization Legacy to implement a bioreserve strategy in this region illustrates the complexity of trying to implement an ambitious and long-term strategy. Education will be the key to successful implementation, for only if people understand both the need for biodiversity conservation and the potential of bioreserves to protect it will progress be made.

This approach to implementation relies heavily upon outreach to the resident citizenry and to various governmental and nongovernmental organizations. Johns and Soulé (1996) have described the approach as follows: "Think of the process as creating an ever expanding circle of people who understand and support wildness and biodiversity: networks of people defending networks of land and water." This philosophy is central to the approach being taken by Legacy in the North Coastal Basin.

Lastly, in the case history described, the proponents of the bioreserve strategy rely heavily upon a bottom-up approach to ultimate resolution of land allocation problems—for example, with watershed groups designing and implementing the building blocks of the bioreserve network. This contrasts somewhat from reports from other regions that appear to rely more on a top-down approach. The latter may be more feasible where much of the landscape remains in public ownership. We believe that the integrated approach described here, with an emphasis on resolving watershed-level issues, is the appropriate one for this ecoregion. As Walters (1996) has written regarding implementation of bioreserve strategies: "[T]o achieve even a measure of political possibility will require . . . acceptance and integration in the context of local struggles. We can never win on a larger stage what we can't win in our own homes and neighborhoods."

Chapter 4

Regional Approaches to Managing and Conserving Biodiversity

J. Michael Scott, Blair Csuti,
R. Gerald Wright, Patrick J. Crist,
and Michael D. Jennings

In a recent article we observed that "[b]iodiversity is largely a matter of real estate. And, as with other real estate, location is everything. . . ." (Kiester et al. 1996:1333). By this we make the point that species, communities, and ecological processes have spatial properties and that their *in situ* conservation, the only conservation strategy that has any hope of maintaining both ecosystem function and evolutionary potential, begins with spatial analysis of distribution information. We originally described the process of Gap Analysis as "[a] geographic approach to protecting future biological diversity" (Scott et al. 1987). Although we focused on the conservation of areas of high species richness in Scott et al. (1987), we also recognized the need for a "systems approach" to conservation planning "while not abandoning the concept of protecting individual endangered species." Over the past decade, the Gap Analysis Program has developed a hierarchical approach of ecosystems (as represented by vegetation cover types), sets of species, and individual species for mapping and analyzing the distribution and status of biodiversity (Csuti and Kiester 1996). Over the past several years, many state Gap Analysis programs have developed digital data layers of land cover, land stewardship (ownership and management), and distribution of terrestrial vertebrates that have allowed a number of regional land-use planning initiatives to consider the effects alternative land-use practices would have on regional biodiversity. Other programs, both in the United States and in Australia, have used Gap Analysis or similar digital data to identify optimal sets of areas in which all land-cover types would be represented. We provide a brief overview of the current status of the Gap Analysis Program and describe the application of Gap Analysis data to regional land-use planning and conservation initiatives.

Although the concept of Gap Analysis was formally described by Burley (1988), the idea of capturing examples of all vegetation types and species in a natural area network is at least a century old. In his 1890 inaugural address to the Australasian Association for the Advancement of Science, F. von Mueller (cited in Ride 1975) said: "Choice areas, and not necessarily very extensive, should be reserved in every great country for some maintenance of the original vegetation, and therewith for the preservation of animal life concomitant to peculiar plants." While von Mueller was writing before the need for natural areas to be large enough to maintain minimum viable populations of animal species was recognized, he clearly called for a natural area network in which all land-cover types were represented, assuming that most animal species would also be represented. This is the first expression of a "coarse-filter" conservation strategy (Noss 1987). In the United States, the Ecological Society of America inventoried the representation of plant communities in natural areas with the goal of identifying further conservation needs (Kendeigh et al. 1950–51). At a global scale, Dasmann (1972) assessed the amount of reserved area in the world's biotic provinces, identifying provinces deficient in reserves. He also recognized the need for finer-scale analysis within each province (Dasmann 1972:251). Specht et al. (1974) assessed the representation of over 900 plant communities in Australian nature reserves. Terborgh and Winter (1983) were among the first biologists to use species distribution maps and centers of richness for endemic species to identify areas with high conservation priority at a national scale (Colombia and Ecuador). The basic concepts of regional conservation planning, using a hierarchical coarse- and fine-filter approach to identify plant communities and animal species not represented in the existing natural area network, are therefore firmly rooted in historical approaches to conservation inventory and planning. The Gap Analysis Program has applied two recent technologies, geographic information systems (GIS) and satellite remote sensing, to mapping the distribution and assessing the status of elements of biodiversity (Scott et al. 1993; Scott et al. 1996; Scott and Csuti 1997a). While these new technologies do not overcome all the problems of mapping the distribution of vegetation cover types or animal species, they do provide a geo-referenced baseline of distributional data suitable for preliminary regional analyses (Edwards et al. 1996) and capable of being upgraded through field surveys and advances in remote sensing technology.

Gap Analysis Data as a Foundation for Regional Planning

Gap Analysis is based on the comparison of maps of land cover and selected animal species with patterns of land ownership and management activities. In practice, there have been no preexisting sources for these three spatial data layers at scales compatible with regional land-use planning. Of necessity, the state and national Gap Analysis programs have developed several new approaches to mesoscale (for example, 1:100,000) mapping of land cover and the predicted distributions of terrestrial vertebrates. A detailed explanation of our current methods is presented by Scott et al. (1996). Below we describe some of the changes in our techniques for building these data layers over the past decade.

Land-Cover Mapping

Land-cover mapping in Idaho, the first state program to use Gap Analysis, was essentially an exercise of creating a 1:500,000 mosaic of existing land-cover maps prepared by the U.S. Forest Service, Bureau of Land Management, and other state and federal agencies (Caicco et al. 1995). The resulting map was edited to incorporate recent land-cover changes by comparison with LANDSAT Multi-Spectral Scanner (MSS) 1:250,000 false-color infrared positive prints. The second Gap Analysis state vegetation map, for the state of Oregon, was based on photo-interpretation of similar LANDSAT MSS 1:250,000 prints. Existing vegetation maps at a variety of scales, produced by various state and federal agencies, were used to help label zones of homogeneous vegetation commonly referred to as polygons. By 1990, programs in California and Utah took a major step forward by developing land-cover maps based on digital LAND-

SAT Thematic Mapper (TM) imagery, with a pixel resolution of 30 × 30 m.

The cost of LANDSAT TM imagery has been a major consideration for Gap Analysis projects in large states. As a result, many partnerships were formed with local offices of state and federal agencies in the interest of sharing the cost of statewide LANDSAT TM coverage. These partnerships were often instrumental in building consensus about the value of GIS data among resource management agencies.

A second major step forward occurred when the U.S. Environmental Protection Agency, the National Biological Service, the U.S. Geological Survey, and the National Oceanic and Atmospheric Administration became partners in the Multi-Resolution Land Characteristics Consortium (Loveland and Shaw 1996). Through their combined resources, complete national LANDSAT TM imagery was purchased for dates with favorable cloud-cover conditions from 1991 to 1993. For some regions, both early-summer and fall imagery is available. These scenes are preprocessed and archived by the U.S. Geological Survey's EROS Data Center in Sioux Falls, South Dakota, and are made available to participating state programs and cooperators.

A final challenge in land-cover mapping has been the consistent labeling of independently developed maps. Orians (1993:206) pointed out that "there is no previously established, generally accepted taxonomy of habitats, communities, or ecosystems." To address this issue, the Gap Analysis Program has adopted a standardized, hierarchical classification for terrestrial land cover patterned after UNESCO (1973) and Driscoll et al. (1984). We have been cooperating with The Nature Conservancy (TNC), the Ecological Society of America, and the federal Geographic Data Committee to adopt this approach to terrestrial land-cover classification as a national standard. Regional TNC ecologists, with Gap Analysis Program support, have been listing plant classes, subclasses, groups, formations, and community alliances in each of their regions. These alliances are the target level for state vegetation cover-type map labeling. We recognize, however, that many valid community alliances exist at scales too small to be captured by any state, regional, or national land-cover mapping effort. Therefore, Gap Analysis Program land-cover maps depict the location, composition, and area only for those community alliances that can be resolved and identified through satellite remote sensing and other mesoscale ancillary data. Where higher-resolution data on the location of occurrences of land-cover types smaller than the minimum mapping unit are available, they can be captured as polygon attributes.

Predicted Distributions of Terrestrial Vertebrate Species

Ideally, a program designed to assess the distribution and status of biodiversity would include all ecosystems, plants, and animals. Because the distribution of many species groups has not been sufficiently documented, we are limited to including the distribution of those groups of organisms for which good distributional data are available. In most regions of North America, we have adequate distributional data only for vertebrates and, occasionally, butterflies or other groups of invertebrates. We feel that, to some degree, vertebrates and land-cover types can serve as a useful surrogate for other unrelated groups of species. While areas of richness for different classes of organisms do not necessarily coincide (Prendergast et al. 1993; Lawton et al. 1994), sets of complementary areas in which all vertebrates are represented also tend to do a reasonably good job of capturing most species of unrelated organisms (Csuti et al. 1997; Kiester, personal communication). Scott et al. (1993), Butterfield et al. (1994), and Csuti (1996) presented overviews of the GAP approach to predicting the distribution of vertebrate species.

Because fish are restricted to aquatic habitats and their distribution is governed by different factors than terrestrial vertebrates, we have incorporated fish and other aquatic organisms into our data sets only in two state pilot programs. As funding becomes available, aquatic Gap Analysis will be conducted nationwide.

Gap Analysis vertebrate distribution maps were developed from mesoscale land-cover maps and other data and are intended to be used and validated at that scale. It would be inappropriate to expect regional maps to predict stand-level species compositions with high accuracy. There have now been several studies aimed at testing the accuracy of GAP predicted species distribution maps. Edwards et al. (1996) compared predicted and observed species lists for eight national park units in the state of Utah. The predictive accuracy of GAP distribution maps varied among vertebrate class, ranging from a high of 90.6 percent for birds (1.86 percent omission error, 7.51 percent commission error) to a low of 69.4 percent for amphibians

(16.07 percent omission error, 14.51 percent commission error). This result is not unexpected because birds interact more closely with terrestrial land cover than amphibians. We have always emphasized the need for field verification of any predicted distribution map at the stand or site level (Scott et al. 1993); however, we feel that GAP vertebrate distribution maps have proven themselves sufficiently accurate for preliminary estimation of faunal communities at the regional scale.

Land Ownership and Land Management Status

One of the more contentious issues faced by the Gap Analysis Program has been the categorization of different land ownerships and land uses with respect to their potential contribution to the maintenance of biological diversity. Public lands are subject to a variety of management prescriptions. Given sufficient time, it would be possible to contact each public land manager and obtain detailed information about management objectives. A nearly limitless number of management categories could emerge from this process, making interpretation difficult. For the purposes of analysis, we assign one of four management status categories to land ownership (Scott et al. 1993): managed primarily for populations of native species and natural ecosystem processes (Management Status 1; e.g., most national parks); managed primarily for natural communities but subject to some uses that may degrade those communities (Management Status 2; e.g., wilderness areas); multiple-use public lands that are subject to intensive uses such as timber harvest, mining, and grazing (Management Status 3; e.g., most National Forest Service and Bureau of Land Management lands); and lands used or potentially used for intensive human activities, such as urban areas or agriculture (Management Status 4).

While we recognize that no land ownership and management status categorization will satisfy the needs or opinions of all interested parties, we feel that further subdivision of our four categories is unwise, given the practical constraints involved in reporting the status of each and every vegetation type and species. Led by the New Mexico Gap Analysis Program, we have recently developed a dichotomous key that we hope will lead to greater standardization and repeatability in the categorization of land ownership and management status. There is an ever-limited area that will be managed exclusively for biodiversity (Pressey 1994), and most biodiversity now occurs and will continue to persist on multiple-use public lands (Scott et al. 1990). Comparing the distribution of biodiversity elements and current land management status informs us of the site-specific opportunities available for maintaining biodiversity in a landscape and regional context.

Reporting the Current Status of Land-Cover Types and Vertebrate Species

As defined by Burley (1988), Gap Analysis determines "which elements [of biological diversity] (for example, major ecosystems, vegetation types, habitat types, and species) are unrepresented or poorly represented in the existing system of conservation areas." While it is possible to quantify estimates of the amount of each vegetation cover type or habitat for each species in different land management categories, there is no objective way to "say with accuracy how much land or what percentage of an ecosystem type must be kept in a natural condition to maintain viable populations of a given proportion of the native biota or the ecological processes of an ecosystem" (Noss et al. 1995:3). The question is intractable in part because different vegetation cover types may have very different natural disturbance regimes and, hence, different minimum dynamic areas (Pickett and Thompson 1978). Likewise, the minimum area required to support a viable population of an animal species will vary with the size, mobility, life history attributes, and ecological requirements of that species. The Devils Hole pupfish (*Cyprinodon diabolis*) has persisted for 10,000 years in a spring a few square meters in area, while the several million acres of Greater Yellowstone Ecosystem may not be large enough to support a viable population of grizzly bears (*Ursus arctos*).

The questions How much is enough land to protect a species or vegetation type? and How many examples of a species or vegetation type are required for long-term viability? span the gap from conservation science to public policy (Noss 1996). Inherent in the definition of population viability is a statement of the acceptable level of risk that a given population will become extinct (Shaffer 1981). Because resources for conservation will always be limited, decisions about the amount of money available for natural area purchase or an area of land that will be managed for natural values will always be made in the political arena. The role of the Gap Analysis Program is to provide decision makers with data on the current representation of major land-cover types and selected groups of species in various land management categories. Several state Gap Analysis programs (e.g., Utah, New Mexico, Wyoming, Washington, California) have prepared or shortly will prepare reports detailing those representation statistics. These reports are available in a variety of formats, including CD-ROM (Utah, Edwards et al. 1995), journal articles (Idaho, Caicco et al. 1995; California, Davis et al. 1995), the GAP World Wide Web homepage (www.gap.uidaho.edu/gap), printed documents (Washington), and final state reports (New Mexico, Wyoming). Eventually, all state data layers will be available over the GAP homepage and in CD-ROM format.

The representation of vegetation cover types and habitat for species in different management classes within a state is usually presented in tabular format. The accompanying text serves to interpret the tables and highlight important trends for the user. For example, most of the coastal scrub community in California's southwestern region has been lost to development. Of the remaining 3,908 square kilometers, nearly 90 percent is privately owned and subject to loss (Davis et al. 1995). Only 5 percent of the 22,859-square-kilometer Modoc Plateau Region of California is managed for natural values, while 58 percent is publicly owned multiple-use land managed for timber harvest and grazing (Davis et al., in press). In this same region, the only and largest population of Baker's cypress (*Cupressus bakeri*), a fire-dependent species, may be threatened by current fire management practices. In New Mexico, 11 of 42 natural-land-cover classes had less than 100,000 hectares in areas managed for natural values. These include such restricted classes as the Madrean Lower Montane Conifer Forest, with a total statewide area of 952 hectares, none of which is protected. Also in New Mexico, the predicted range of the Jemez Mountains salamander covered 25,500 hectares, none of which was in Management Status 1, but 20 percent of which (5,150 hectares) was in Management Status 2. A preliminary analysis of land-cover types in the state of Oregon indicated that mountain hemlock (*Tsuga mertensiana*) forests, found in the higher elevations of the Cascade Mountains, had over 50 percent of their area in designated wilderness, whereas a variety of Oregon white oak (*Quercus garryana*) woodlands, found in the lower elevation valleys of western Oregon, had less than 2 percent of their total area in Management Status 1 or 2. These results are indicative of the powerful conclusions that can be drawn from a simple comparison of the distribution of vegetation cover types or species with various land management classes.

Identifying Priority Areas for Conservation

The distributional data assembled by Gap Analysis have a number of land-use planning applications. One of these is the identification of areas of high conservation value. These may contain examples of vegetation cover types or species that are unrepresented or poorly represented in existing natural areas; they may be areas especially rich in biodiversity; they may be essential complements to existing natural areas if all elements of biodiversity are to be represented in natural areas; or they may be the locations of rare or threatened species or species not otherwise represented in natural areas.

A simplistic approach to the identification of sites of high conservation value would be to examine the distribution and conservation status of every vegetation cover type and species individually and to create a conservation plan for each. This approach is inefficient and is likely to identify an impossible number of potential natural areas, especially if multiple sites are identified for each of the hundreds of species present in a state or region. Because the distributional areas of many species overlap, spatial analysis can identify a smaller number of sites in which all species are predicted to occur (Scott and Csuti 1997a). Although examples of different vegetation types by definition do not overlap, different types are contiguous and a large variety of types may occur in close proximity in areas of high environmental variability. Over the past decade, considerable progress has been made in developing quantitative methods for identifying an efficient subset of sites in a region on which all species or land-cover types are thought to be represented once, twice, three times, and so forth (see Pressey et al. 1993 for a review).

When identifying a set of sites for the maintenance of biodiversity, it is important to distinguish between a collection of sites that is especially rich in species and minimal or near-minimum sets of sites in which all species are represented. Csuti et al. (1997) pointed out that the ten areas highest in vertebrate species richness in Oregon are clustered in one part of the state and likely have nearly identical species content. In contrast, the ten sites that most efficiently represent Oregon's vertebrate diversity are widely separated and provide representatives of most of the state's ecological variety. This is the critical difference between selecting a representative set of sites and identifying species-rich areas: sites that are maximally complementary need not be especially rich in species; they merely must contain species not already represented in previous selections (Csuti and Kiester 1996; Williams et al. 1996). This result is a consequence of the principle of complementarity (Pressey et al. 1993) built into most quantitative site selection algorithms. The simplest algorithm is a richness-based heuristic algorithm that first selects the site with the highest species richness. That site is fixed, and the second site selected is the one that adds the greatest number of species not already found at the first site. In turn, subsequent selections add sites with the most species not represented in previous ones. Over 90 percent of all species were represented in the first five sites selected for the Oregon data set (Csuti et al. 1997). In areas with greater local endemism than Oregon, more sites will be needed to achieve this goal.

Underhill (1994) has pointed out that simple heuristic algorithms often fail to find the optimal, or minimum possible, combination of sites that most efficiently represents all species. Ideally, all possible combinations of two, three, four, . . . n sites should be examined to find the solution that optimally "covers the set." These approaches differ from heuristic algorithms in that areas selected when the problem is solved for a smaller number of sites are not fixed when the problem is solved for any larger number of sites. Using a data set from Idaho, it was not possible to compute all possible combinations of more than five sites (Kiester et al. 1996). Underhill (1994) and Camm et al. (1996) suggested that branch-and-bound linear programming algorithms, developed in the field of operations research, provide a way to identify optimal combinations of a larger number of sites. Church et al. (1996), using GAP data from California, and Csuti et al. (1997), using GAP data from Oregon, demonstrated linear programming solutions that cover the set (that is, represent all species) for data sets containing several hundred species and several hundred sites.

While there are intuitive and practical advantages to identifying optimal solutions to reserve selection problems, it may not be possible to calculate them for very large data sets, and computation time may be too long to be practical during interactive meetings with planners and decision makers (Pressey et al. 1996). Both Csuti et al. (1997) and Stokland (1997) have demonstrated that site selection solutions derived from relatively simple heuristic algorithms can be nearly as efficient as optimal solutions and can be calculated in a matter of seconds for large data sets.

We offer four cautions when translating the results of site selection algorithms into conservation recommendations. First, all algorithms select the richest site for a given data set first. The data set may be all vegetation cover types, all species, or all species in a group (such as mammals or neotropical migratory birds). This first choice is easily defended, but it is retained in the solution only by heuristic algorithms. Because sites may or may not be included in linear programming solutions at different cardinalities, their absolute importance to maintaining biodiversity in an area is more difficult to assess. Pressey et al. (1994) and Csuti et al. (1997) suggested that this importance might be expressed as an irreplaceability value—that is, the percentage of all fully representative sets of sites in which an individual site participates.

Second, not all areas of high conservation importance are under the same degree of threat of degradation or destruction. The degree of threat to a site needs to be considered when setting conservation priorities.

Third, most indicators of biodiversity (vegetation cover types, species) will be represented in a relatively small number of sites. The last sites selected tend to add only one or two species per site. It seems unrealistic to consider these sites as important to the maintenance of biodiversity as especially rich sites. The species driving the selection of the last several sites are often rare in the study area, either because they are intrinsically rare or because they are peripheral to the region. Their conservation needs may best be considered through a fine-filter strategy.

Fourth, the importance of any site identified as a possible conservation priority through the application of reserve selection algorithms to a spatial data set must be field validated. If the area is confirmed as supporting high-quality habitat and viable populations of its resident species, then other ecological parameters must be employed to determine the required size and shape for natural area. Gap Analysis provides the spatial information needed to identify potential priority areas for conservation; selecting among alternative potential priority areas and designating natural area boundaries are the next steps in implementing a regional conservation strategy (Scott and Csuti 1997b).

Regional Land-Use Planning to Maintain Biodiversity

The production and dissemination of mesoscale digital data on the distribution of land-cover types and vertebrate species invite their application in a variety of regional land-use planning processes.

- Wright et al. (1994) used Gap Analysis land-cover and land ownership data layers to evaluate the potential contribution of four areas (average size, 220,000 hectares), proposed as new national parks or additions to existing national parks in Idaho, to the protection of regional vegetation cover types. They found that the proposed additions did not significantly increase the level of protection afforded to land-cover types already represented in natural areas. However, relatively small boundary changes in the proposed additions offer the potential to add to the protection status of many land-cover types, pointing out the importance of considering regional land-cover distributional information in the park planning process.

- At the request of the governor, the Utah Gap Analysis Program analyzed the contribution that alternative proposals for increasing the amount of wilderness in Utah would make toward the goal of representing the state's biodiversity in protected areas. Contrary to intuition, the proposal to add the largest amount of new wilderness did not contribute most to representing land-cover types or vertebrate species in those areas. This outcome reflected the fact that few land-cover types not already represented in wilderness were included in the proposed additions. These same data were considered in evaluating the contribution of the new Escalante-Grand Staircase National Monument to the conservation of Utah's land-cover types and vertebrate species.

- Steinitz et al. (1996) carried out an intensive analysis of the effects that six different alternative future land-use models would have on the physical, biological, and cultural environment of a 10,729-square-kilometer area of southern California surrounding the Camp Pendleton Marine Corps Base. Their study drew on a number of Gap Analysis/ Biodiversity Research Consortium data sets for basic land-cover and vertebrate species information. These data were supplemented by a wide variety of data describing the topography, soils, fire history, hydrology, and human use, both current and planned, of the region. While all alternatives are designed to account for projected future population growth within the study area, they have significantly different outcomes for the biological resources, hydrology, and fire regimes in the region. Although no specific alternative future is recommended, this study provides regional planners with a far clearer picture of the costs and benefits of different patterns of future development and habitat conservation.

- The USDA Forest Service and the USDI Bureau of Land Management recently completed a major study of over 58 million hectares in the Columbia River Basin, intended to provide a framework for future ecosystem management within the region for both federal agencies (USDA Forest Service 1996). Gap Analysis land-cover maps and vertebrate distributional data from the states of Washington, Oregon, and Idaho contributed to the data used in this study. These and other data were developed by the study team into maps of "Current Potential Vegetation Groups" and "Hot Spots of Species Rarity and Endemism, and Hot Spots of Biodiversity." According to the team's report, "The species-environment relations database provides a means of identifying such species by vegetation type, as well as a means of prioritizing species for study" (USDA Forest Service 1996:100). One potential outcome of this study is the "realignment" of natural areas to "better coincide with hotspots of species rarity and endemism and hot spots of high biodiversity" (USDA Forest Service 1996:100).

- The Defenders of Wildlife is spearheading a three-year effort to develop and implement a statewide biodiversity conservation strategy for Oregon, known as the "Oregon Biodiversity Project." Here researchers are using Gap Analysis actual vegetation types as starting points for a detailed analysis of the state by ecoregion and ten-digit U.S. Geological Survey hydrologic units. The project has added over fifty GIS data layers to the data set for their analyses, including information on geothermal and mineral sites, road densities, grazing allotment condition, soils, census data, and election results on natural resource issues. These findings (Oregon Biodiversity Project 1998) will be the foundation for a series of initiatives for regional biodiversity conservation planning.

- The New South Wales National Parks and Wildlife Service has carried out regional analyses to identify areas that best represent examples of all remaining old-growth eucalyptus forest types. These studies use sophisticated, interactive software programs developed by Robert L. Pressey and his colleagues to display conservation and harvest options to both forestry and conservation stakeholders. Through intensive workshops lasting several days, all parties are brought to agreement on those areas that should be deferred, at least pending further study, from timber harvest. This has resulted in the short-term regional protection of over 800,000 hectares of old-growth forest in New South Wales (R. L. Pressey, personal communication).

- The southwestern ecoregion of California, as defined by Hickman (1993), includes 8 percent of the area of California but over half its human population (nearly 20 million people). This concentration of human activity in a relatively small area has resulted in considerable loss of native plant communities, especially at lower elevations (Davis et al. 1995). In turn, populations of many terrestrial vertebrate species have declined to the point of concern, and several are federally listed as endangered (Stephens' kangaroo rat [*Dipodomys stephensi*], least Bell's vireo [*Vireo bellii pusillus*], California coastal gnatcatcher [*Polioptila californica californica*]). Stine et al. (1996) tested the ability of mesoscale land-cover maps to identify remaining coastal sage scrub habitat in comparison to independently developed fine-scale (1:24,000) vegetation cover maps. Not unexpectedly, while finer-scale maps located more small polygons of this habitat type, satellite-based GAP maps located nearly all stands large enough to make meaningful contributions to maintenance of the region's biota (larger than 100 hectares). Church et al. (1996) carried out a Gap Analysis of the southwestern ecoregion of California and identified twelve sites in which all 333 terrestrial vertebrates of the ecoregion are predicted to be represented. As was observed in Oregon (Csuti et al. 1997), over 90 percent of all species could be represented in only five sites.

- Davis and Stoms (1996) and Davis et al. (1996) have also applied Gap Analysis data to land-use planning in the Sierra Nevada ecoregion of California. They found both subregional and elevational unevenness in the representation of vegetation cover types in existing reserves. Nearly all oak woodland is found in private ownership, and nearly all is subject to grazing. In contrast, a majority of high-elevation pine forest types tend to be well represented in reserves. These results mirror those found in a preliminary analysis of western Oregon vegetation cover types. Davis et al. (1996) also identified the locations of smaller watersheds (about 4,000 hectares in size) in which both 10 percent and 25 percent of each vegetation cover type would be represented. They found that watersheds in existing national parks and wilderness areas do not capture the range of vegetation diversity in the region, but that a similar total percentage of the region would be sufficient to capture representative examples of all vegetation types, although that percentage would have to consist of more numerous and more evenly dispersed watersheds.

- The Gap Analysis Program has long recognized the need to assess the conservation status and needs of elements of biodiversity in a bioregional rather than state context (Scott et al. 1993). Consistent bioregional maps of vegetation cover types and vertebrate species are required for regional analysis. Recently, a number of GAP projects in contiguous states have completed vegetation mapping, facilitating the first Gap Analysis of an entire region, the Intermountain Semi-Desert Ecoregion (as delineated by Bailey 1995). This analysis has been carried out under the direction of David Stoms, Frank Davis, and others at the Institute for Computational Earth System Science, University of California, Santa Barbara (Stoms et al. 1998). The distribution of forty-eight vegetation alliances was mapped over the 412,000-square-kilometer region. Over 60 percent of the land in this region is federally owned, but less than 4 percent is categorized in Management Status 1 or 2. Two vegetation alliances (Jeffrey pine and alpine tundra) were not represented in any protected area, and an additional seven alliances had less than 1 percent of their total extent in protected areas. Overall, twenty-two alliances were deemed vulnerable to potential loss or degradation. This ecoregion is also the subject of a pilot ecoregional planning exercise by The Nature Conservancy. The combination of traditional fine-filter methods, refined by The Nature Conservancy's Natural Heritage/Conservation Data Center network, with a coarse-filter overview based on regional Gap Analysis land-cover mapping, holds the promise of developing integrated ecoregional conservation plans based on multiple levels of the biodiversity hierarchy.

Summary and Conclusions

Regional land-use planning has long been handicapped by a lack of consistent, high-quality mesoscale (1:100,000) information about the distribution of land cover and other natural resources. The Gap Analysis Program has developed these data and made them available to a variety of planning initiatives. As a result, a variety of planning agencies have been able to determine the location of biologically significant areas within a region and to assess the outcome of different land-use alternatives on both public and private lands. This application of Gap Analysis data to regional planning provides us the opportunity, as demonstrated by the Habitat Conservation Planning effort in the southwestern ecoregion of California, to anticipate and avoid future conflicts between conservation and development, averting ecological "train wrecks" through proactive planning rather than reactive regulation.

Chapter 5

Application of Historical Range of Variability Concepts to Biodiversity Conservation

Gregory H. Aplet and
William S. Keeton

In the age of ecosystem management, no single theme has dominated the conservation of biological diversity more than the role
of disturbance in maintaining ecosystem composition, structure,
and function. The relationship between ecosystem complexity and
dynamic processes is not an entirely new topic in ecology; succession, the changes that occur in ecological communities after disturbance, has been a central focus of ecology since the advent of
the science (Clements 1916; Watt 1947), and the role that fire plays
in maintaining natural ecosystems has been acknowledged for
decades (see review by Parsons 1981). Nevertheless, the ubiquity
of disturbance as an organizing force in ecosystems has only recently become widely appreciated (Pickett and White 1985; Botkin
1990). It is now understood that ecosystem conditions change
over time as they are affected by disturbance; when disturbances act
with a characteristic behavior, ecosystems exhibit a characteristic
behavior and complexity. When humans act to alter disturbance patterns, such as through traditional logging or fire suppression, ecosystems change. As Pickett et al. (1992) noted, "Human-generated
changes must be constrained because nature has functional, historical, and evolutionary limits. Nature has a range of ways to be, but
there is a limit to those ways, and therefore human changes must
be within those limits." The attempt to understand and apply these

limits to sustain the complexity and dynamics of ecosystems is the subject of this chapter.

What Is Historical Range of Variability?

All ecosystems may be described in terms of composition, structure, and function (Landres et al. 1998). "Composition" refers to the abundance, or relative abundance, of components, such as water, nutrients, and species, that make up the ecosystem. "Structure" refers to their physical arrangement in space, and "function" refers to the processes through which composition and structure interact, including predation, decomposition, and disturbances, such as fire and floods. Because ecosystems are dynamic, these attributes are constantly changing, but composition, structure, and function are constrained within limits. The bounded behavior of an ecosystem can be called its "range of variability."

Figure 5.1 provides a simplistic illustration of range of variability, a hypothetical fluctuation of old-growth acreage within a watershed. Over time, old-growth acreage increases as stands mature until a large, catastrophic disturbance event, such as fire, consumes some old growth and the acreage declines. The dynamics of old growth in this system describe the range of variability of this ecosystem component.

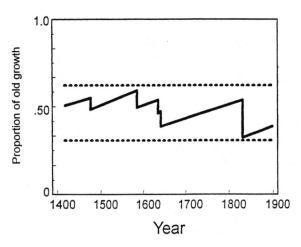

Figure 5.1. Graphical representation of historical range of variability showing fluctuation in the area of old growth in a hypothetical watershed.

The phrase "range of variability" has been justly criticized for failing to communicate important aspects of ecosystem behavior (Rhodes et al. 1994; Frissell and Bayles 1996), and some authors have concerned themselves only with the range of, or difference between, high and low values of ecosystem attributes (Caraher et al. 1992). However, most authors who have explored the meaning and utility of the concept have agreed with Morgan et al. (1994) that "the rate of change in ecosystem characteristics is as important to the concept of historical range of variability as is the magnitude of historical fluctuations." In other words, the phrase "range of variability" conveys much more than extreme values; it also describes rates of change and duration of states represented by the fluctuations (Swanson et al. 1994; Morgan et al. 1994; Landres et al. 1998). The range of variability is thus a portrait of the dynamic behavior of an ecosystem.

The range of variability concept can also be thought of as a dynamic version of a well-established model in ecology. Over a forty-year period, Jenny (1941, 1980) refined a model of ecosystem development that he called the "state factor model." According to the state factor model, the state of any ecosystem, e, is a function of a relatively small number of "state factors," including local climate (cl), the organisms available to colonize the site (o), topographic relief (r), soil parent material (p), time since the last disturbance (t), and additional, unspecified factors. (Jenny represented this model as $e = f(cl, o, r, p, t \ldots)$, thus, the model is sometimes known as the "clorpt model.") Recent developments in disturbance ecology have shown that other factors besides time since the last disturbance are important (White and Pickett 1985), so t is probably better thought of as describing a multifactor "disturbance regime." In dynamic ecosystems, the behavior of ecosystems should remain bounded as long as the state factors controlling ecosystem behavior are themselves bounded.

"Historical range of variability" (HRV) describes the bounded behavior of ecosystems prior to the dramatic changes in state factors that accompanied the settlement of North America, beginning with the discovery of the "New World." For several thousand years prior to colonization, topographic relief and parent material remained relatively constant. Climate fluctuated but narrowly, relative to the Quaternary or even the entire Holocene (Johnson et al. 1994; Woolfenden 1996). Rates of extinction and species arrival were relatively low (OTA 1993), and disturbance operated according to characteristic regimes (Mooney et al. 1981). Population levels and technologies of indigenous people may have had widespread ecological impacts, which were severe in some localized cases (Betancourt and Van Devender 1981), but as those impacts were relatively consistent across the landscape over several thousand years (see, for example, Anderson and Moratto 1996). Resulting ecosystems were likewise bounded in their behavior; they maintained characteristic species composition, pattern, and behavior.

Since European settlement, many ecosystems have changed dramatically as a result of changes in state factors. The organism state factor has been altered as some species have become extinct and many others have been introduced from other continents (OTA 1993). The disturbance state factor has been even more dramatically changed. Through forest clearing, agriculture, timber harvest, fire suppression, and other effects, the composition, structure, and function of entire regions have changed (Thomas 1956; Whitney 1994). While Native Americans manipulated the landscape for thousands of years, their technologies were far different from those employed by European settlers (Cronon 1983; Anderson and Moratto 1996). The recent colonization of North America may be thought of not as a biological invasion, but as a technological invasion—with devastating effects on biodiversity and ecosystem integrity. The concept of HRV is useful in describing ecosystem dynamics prior to these dramatic changes—the ecosystem dynamics that had sustained biodiversity and ecosystem integrity over the thousands of years before the devastating changes of the last few hundred years.

This bounded ecosystem behavior goes by a number of different names, including "range of natural variability" (Caraher et al. 1992; Landres et al. 1998), "natural variability" (Swanson et al. 1994), and "reference variability" (Manley et al. 1995). "Historical range of variability" differs from the others by conveying a sense of time, an essential quality of the concept (Morgan et al. 1994), and it avoids use of the term "natural," which has been criticized because of the ubiquity of human impacts in ecosystems, the changing nature of ecosystems over time, and the imprecision of descriptions of communities and ecosystems in ecological literature (Schrader-Frechette and McCoy 1995; Hunter 1996). By explicitly acknowledging the time frame of observation and the role of humans in shaping ecosystems, HRV can provide an approximate description of "natural conditions" that addresses these criticisms (Comer 1997). Though we prefer the use of "historical," we intend that HRV be interpreted as a description of "natural conditions."

The assumption behind the use of HRV in management is that "restoring and maintaining landscape conditions within distributions that organisms have adapted to over evolutionary time is the management approach most likely to produce sustainable ecosystems" (Manley et al. 1995). According to this approach, if landscapes can be maintained within HRV, or a socially preferred range based on HRV, those landscapes stand a good chance of maintaining their biodiversity and integrity over time. Sustainable management employs historical information as a reference for restoring and maintaining the patterns and processes characteristic of North American landscapes prior to recent alteration.

Of course, no management approach can be expected to provide a "silver bullet," and HRV has sustained its share of criticism. Obtaining historical information for many important aspects of ecosystems is difficult, variation in ecosystem conditions (especially climate) limits its usefulness, HRV is scale dependent, and presettlement conditions have been criticized as an arbitrary model for the future (Rhodes et al. 1994; Hunter 1996; Millar 1997). Nevertheless, historical information can be very helpful in understanding and illustrating the dynamic nature of ecosystems, the processes that sustain and change ecosystems, the current state of the system in relation to the past, and the possible ranges of conditions that are feasible to maintain (Morgan et al. 1994). To ignore historical information because of the difficulties it presents "would be to throw the baby out with the bathwater" (Millar 1997).

In theory, there is no limit to the number of ecosystem variables that can be used to describe HRV. HRV can describe changes in soil conditions, animal population sizes, composition of the plant community, stream sediment loads, air quality, human communities, or any other aspect of ecosystems that can be quantified. Indicators of ecosystem condition can be characterized with respect to HRV to help understand management impacts and preferred directions for the future. Manley et al. (1995) compiled over three dozen "key ecosystem elements" for which historical ranges can guide management. In practice, however, historical information is very difficult to obtain for most aspects of ecosystems (Rhodes et al. 1994; Millar 1997). In most cases, HRV should be limited to those few aspects of ecosystem condition for which historical information is available or can be confidently inferred. Some of the best sources of historical information include tree rings, fire scars, and pollen cores, which can divulge information about species composition and disturbance regimes hundreds or even thousands of years ago. If changes in conditions can be understood in terms of processes, historical information can be used in conjunction with simulation models to infer HRV (Baker 1992), though investigators must be careful not to "pile inference upon inference" (Millar 1997).

In general, the ecosystem attribute for which we seem to have the most historical information is vegetation, particularly forests; thus, HRV is often described, appropriately, in terms of forest composition (e.g., hectares of old growth), structure (e.g., patch size distribution), and function (e.g., fire frequency). While there is much more to the description of ecosystem composition, structure, and function than vegetation, this one feature can provide insights into a broad range of ecosystem attributes, including wildlife habitat, aquatic systems, visitor experience, and fire behavior. As Manley et al. (1995) noted, "The vegetation mosaic is a Key Ecosystem Element when determining habitat suitability for all biological species and individuals."

Application of HRV to the Conservation of Biodiversity

The objective behind the application of HRV to biodiversity conservation is to sustain into the future the ecosystem conditions that sustained biodiversity prior to the dramatic changes of the recent past. Fundamental to this approach is the concept of "representation" (Noss and Cooperrider 1994), which aims to maintain on the landscape conditions that represent all of the variety extant in nature. Historically, representation has been used to help identify lands (for example, vegetation types) that should be included in a conservation reserve system (see Chapters 2 and 3). This is the same approach to conservation advocated in the 1920s by the Committee on the Preservation of Natural Conditions of the Ecological Society of America (Shelford 1926) and later by The Nature Conservancy (1982). Representation was also used by the USDA Forest Service in its evaluation of the wilderness potential of roadless areas (RARE II) in the 1970s (Foreman 1995–96). Alternatively, representation can be achieved by maintaining the historical range of variability across the landscape, both in reserved and nonreserved lands.

Application of HRV to Improve Coarse-Filter Conservation

As a basis for determination of a reserve system, representation is certainly a key concept, but it is insufficient. Representation address-es only the question of what to protect, not how much or in what arrangement or how to sustain it in a dynamic ecosystem. Under-standing the historical ecosystem behavior implicit in the concept of HRV can help improve reserve-based conservation by providing insights into the kinds and amounts of ecological elements that should be included in a coarse-filter reserve system (see Chapter 2) and what processes are necessary to sustain them into the future.

Because both HRV and coarse-filter conservation generally involve descriptions of vegetation, insights derived from HRV analysis can be readily applied to improve the coarse-filter approach. For example, coarse filters have been criticized for employing vegetation classifica-tion systems that fail to account for the dynamic nature of vegetation (Hunter 1991). Analysis of HRV can help identify dynamic stages of vegetation within vegetation types, and coarse-filter protection can be applied to those dynamic stages. Also, coarse filters are tradition-ally applied to existing vegetation evenly, without regard to the his-torical importance of vegetation classes. By describing the bounded abundance of vegetation types and seral stages, HRV can help iden-tify the appropriate amount of each vegetation class that should be included in a reserve, and it can help identify for protection those vegetation elements that were important historically but that are rare or absent today (Haufler 1994).

Other aspects of reserve design not adequately addressed by tradi-tional coarse-filter approaches are habitat arrangement and patch size. Representation fails to account for the juxtaposition of vegeta-tion patches and its influence on ecosystem function (Chapel et al. 1995). For example, species that forage at the boundary between two vegetation types will find no suitable habitat if the two types are pro-tected on separate tracts of the coarse-filter reserve system. Also, dis-tance between patches can affect dispersal and colonization and limit reserve effectiveness (Soulé and Simberloff 1986). Patch *size* is also an important variable determining the suitability of habitat for many species (Robbins et al. 1989; Thomas et al. 1990). The effectiveness of a reserve system may be improved if vegetation patch size and arrangement are informed by historical analysis of the landscape. Reconstruction of historical patch dynamics (see, for example, Morrison and Swanson 1990) can reveal insights into these important variables. Ultimately, the goal of reserve design is to identify an area in which natural disturbance processes will maintain these characteristics over time.

The application of historical dynamics to reserve design is not entirely new. Pickett and Thompson (1978) introduced the phrase "minimum dynamic area" to describe "the smallest area with a natural disturbance regime, which maintains internal recolonization sources, and hence minimizes extinction." These authors recognized that disturbances would inevitably change the very conditions for which a reserve was established unless the reserve were large enough to "absorb" characteristic disturbances. Such a reserve would be in a state of "dynamic equilibrium"; the proportion of vegetation patches would remain constant, though the location of patches in various stages of development would change over time (Shugart and West 1981). The advantage of setting reserve size above the minimum dynamic area based on HRV is that natural disturbance can continually refresh the appropriate amount of habitat and maintain appropriate patch size and juxtaposition, provided that the disturbance regime has not been fundamentally altered.

Application of HRV to Achieve Representation across the Landscape

To summarize the above discussion, the coarse-filter approach to conservation attempts to sustain biodiversity by protecting examples of plant communities over time. These reserves must be large enough to sustain viable populations of constituent species and be capable of sustaining their character in the face of disturbance. Where species have minimal habitat requirements and disturbances are small and infrequent, this can be accomplished on relatively small areas. Where territories and disturbances are large, reserves must be carefully planned to provide adequate habitat and endure inevitable disturbance.

It is increasingly clear, however, that even the very largest reserves are insufficient to meet the conservation requirements of a coarse-filter system. Populations of grizzly bears, bison, and elk rely upon the millions of hectares of wilderness and roadless areas outside Yellowstone National Park to meet their habitat needs (Harting and Glick 1994). Eight hundred and ten thousand hectares were not enough to protect half of the reserve from being consumed by fire in 1988. The conservation of late-successional forest ecosystems in the Pacific Northwest required approximately 3 million hectares of reserves distributed around the region, but even these reserves were insufficient to conserve the ecosystem without riparian reserves and management guidelines for the intervening lands (FEMAT 1993). It is now also clear that conservation cannot be accomplished on reserves alone; reserves must be integrated with management guidance on the rest of the landscape to stitch together a functional landscape ecosystem.

Analysis of HRV can help identify the elements of biodiversity to be protected in reserves and can help develop prescriptions for intervening lands that will sustain the landscape. Instead of protecting some of everything, the coarse filter should sift out those rare, valuable, or slowly changing portions of the landscape ecosystem identified through analysis of historical ecosystem behavior. As an example, old growth may exist in places that are refugia from fire (Camp 1995); these settings can be identified for protection in reserves. In addition to plant communities, rare elements of landscape diversity, such as roadless areas, and biodiversity hotspots are worthy candidates for protection even if they are relatively dynamic. Large blocks of land with high ecological integrity are likewise outstanding candidates for inclusion in a coarse-filter reserve network as they retain the highest likelihood of continued integrity in the absence of human intervention. Protecting large blocks with high integrity and smaller tracts supporting key elements of biodiversity recognizes the hierarchical nature of ecosystems by employing a structured system that spans multiple spatial scales (Sullivan and Shaffer 1975; Noss and Cooperrider 1994).

On lands outside of reserves, HRV can help identify the communities, patterns, and processes necessary to sustain the integrity of the landscape ecosystem. It is on these "matrix" lands that HRV has its greatest potential as a model for sustainable management. Franklin (1993) identified three critical roles for nonreserve matrix lands in conserving biodiversity: (1) providing habitat at smaller spatial scales, (2) increasing the effectiveness (buffering) of reserved areas, and (3) providing for connectivity. All three of these roles can be enhanced by application of historical information to matrix management.

Just as an understanding of historical ecosystem behavior can be used to guide the identification of reserves, that same information can be used to derive objectives for vegetation management in the matrix. Under this approach, the overarching goal of matrix management would have to change from maximizing production to sustaining biodiversity. Production of fiber and forage would remain appropriate uses of the matrix, but the objective of management would be to sustain historically appropriate seral patches across the landscape.

Managing the matrix as a dynamic mosaic allows for both the harvesting of materials and the production of key habitats away from reserves. It also addresses one of the principal shortcomings of reserve-based biodiversity conservation: the inevitable loss of habitat from disturbance (Keeton and Aplet 1997). Maintaining HRV in the matrix ensures the availability of a continuous supply of replacement habitat to replenish that lost from reserves (provided that both reserves and matrix lands have the same potential to provide habitat).

Managing the matrix within HRV also extends the effectiveness of biodiversity reserves. As already discussed, many reserves are of insufficient size to conserve viable populations of constituent species. Managing the landscape within HRV effectively increases the size of the landscape that can support native biodiversity. Advocates of the "core reserve approach" to conserving biodiversity (Noss and Cooperrider 1994; DellaSala et al. 1996) often describe a "multiple-use buffer zone" that is managed to increase the effective size of a reserve. Managing the matrix to sustain historical ecosystem conditions, rather than for maximum production, would serve the purpose of extending the effectiveness of reserves without requiring the explicit designation of a buffer zone (Franklin 1993).

Finally, maintaining the matrix within HRV would provide a level of connectivity across the landscape consistent with historical ecosystem function and the needs of constituent species. Many reserve-based conservation strategies rely on habitat corridors to maintain connectivity across the landscape (Noss and Cooperrider 1994). However, the effectiveness of corridors has been challenged because of a lack of evidence that itinerant or dispersing species will actually use them (Simberloff et al. 1992). Maintaining the landscape mosaic within historical conditions would provide the habitat distribution similar to that which dispersing species historically encountered, rather than requiring them to follow predetermined habitat corridors.

It should be noted that applying HRV to land management in the absence of reserves is not advocated here. In a perfect world, the landscape might have been managed to maintain historical conditions without interim protection for lands of high ecological integrity. Unfortunately, the world we have inherited has suffered decades to centuries of human impacts that now threaten its integrity. Restoring healthy ecosystems is not simply a matter of managing within the "free space" (Frissell and Bayles 1996) or "slop" (Aplet et al. 1993) that we hope exists in natural ecosystems. Instead, we must protect ecosystem integrity where it remains high and work on the degraded lands to restore HRV to the whole landscape (Keeton and Aplet 1997). Only once ecosystem health has been restored to the broader landscape can we turn our attention toward harvesting the redundancy or "free space" in ecosystems. As Frissell and Bayles (1996) have said, "[I]f there ever was a free lunch, we already ate it."

Viewing the landscape as an integrated complex of reserve and matrix, with each designed and managed to sustain historical pattern and process, requires a change in the way we think about the land. Traditionally, wilderness areas have been "set aside" and protected for natural values. By implication, lands outside wilderness have been the object of management and are not meant to be natural. But ecosystem research is showing that ecosystems do not end at the borders of reserves. Sustaining ecosystems requires attention to the whole landscape. It is this integrated, whole landscape of reserves and matrix lands that must become the object of management (Franklin 1993). Reserves, including wilderness, can no longer be thought of as "protected from management." Management practices in and out of reserves will differ, but the land must be seen as one.

Barriers to the Application of HRV

Unfortunately, as is so often the case in natural resource management, what appears simple in concept is quite complex in practice. One factor complicating application of HRV is how to translate fluctuation into a management objective. Research in disturbance-prone ecosystems has demonstrated that variability itself can be important to the life histories of many organisms. Managing ecosystem attributes at a constant level, even if it is within HRV, fails to produce the heterogeneity important to ecosystem function (Christensen 1988; Baker 1992; Rhodes et al. 1994). For instance, short-term fluctuations in habitat availability and corresponding fluctuations in population sizes do not necessarily jeopardize species or populations in a dynamic system; however, long-term persistence of population sizes at low historical thresholds can create problems through inbreeding, genetic drift, and increased susceptibility to random environmental and demographic events (Shaffer 1981). Christensen (1988) concluded, "In many ecosystems, one of the most important consequences of fire (or natural disturbance in general) is to maintain landscape heterogeneity. The simulation and maintenance of that heterogeneity is one of the most significant challenges to natural landscape managers."

Aquatic systems present a special challenge in this regard. Riparian and aquatic ecosystems are highly dependent on disturbance for rejuvenation and inputs of resources. Often, natural disturbances are quite violent, exceeding the perturbations caused by humans (Frissell and Bayles 1996). Instead, human impacts tend to be chronic, arising "from the cumulative and persistent effects of thousands of miles of roads, thousands of dams, and century of logging, grazing, mining, cropland farming, channelization, and irrigation diversion" (Frissell and Bayles 1996). Simply managing "within the range of natural variability" will not maintain the temporal heterogeneity necessary to sustain the natural function of these ecosystems. The 1996 release of floodwaters from Glen Canyon Dam through the Grand Canyon was an explicit attempt to reintroduce a dramatic perturbation to a disturbance-dependent ecosystem that had experienced too little fluctuation (Wegner 1996). Developing similar prescriptions to sustain disturbance processes and heterogeneity presents a tremendous challenge.

Another factor complicating application of HRV is scale (Morgan et al. 1994). It is now widely recognized that ecosystems occur at a variety of spatial scales, with smaller systems nested in larger systems in a hierarchical fashion. Properties of large ecosystems are the product of the properties of smaller constituent ecosystems. For example, the forest vegetation of the central Rockies may be thought of as occurring in three zones: ponderosa pine at the low elevations, lodgepole pine at mid-elevations, and spruce-fir forest in the subalpine zone. At a smaller scale, say the subalpine zone of Rocky Mountain National Park, one finds that lodgepole and spruce-fir forests are often found on adjacent slopes, depending on aspect. At the scale of a small watershed, vegetation may vary dramatically, depending on time since catastrophic fire. Different processes operate at each scale to produce the observed patterns. Observations of ecosystem phenomena are thus scale dependent. How we characterize pattern is a function of the scale of observation.

HRV is no different. Observed fluctuations in, for instance, old-growth acreage are going to be very different, depending on whether that observation is made at the scale of a watershed, a national forest, or a region (Figure 5.2). In a small watershed subject to large distur-bance, old-growth acreage may fluctuate widely but infrequently, as

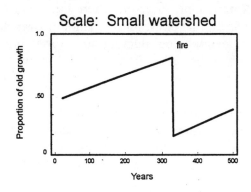

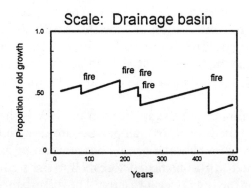

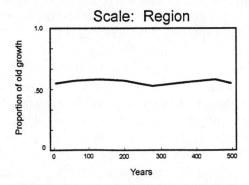

Figure 5.2. The historical range of variability for any given variable is a function of the scale of observation.

in the case of the rare catastrophic fire killing trees over a large area. However, at the scale of the national forest, fire resulting in the loss of old growth somewhere in the forest is a more frequent phenomenon, ensuring that old growth remains somewhere on the forest but never at the high proportion occurring in a single watershed. At the scale of the region, catastrophic fire somewhere in the region is virtually guaranteed on an annual basis, but the amount is small relative to the total, thereby dampening fluctuations. Variability in old-growth acreage reflects cyclic variability in regional fire weather (climate). Thus, HRV is very different, depending on the scale of observation.

The issue of scale dependence of HRV is important because misapplication of the concept to management may lead to inappropriate conclusions. For example, a watershed composed mostly of old growth may be targeted for harvest because it is "outside of HRV," according to the estimate of old-growth HRV derived at the regional scale. Such misapplication might lead to further loss of old growth from a region already drastically below HRV. The proper application of HRV would lead to conservation of watersheds with abundant old growth until old-growth HRV is restored at the landscape or regional scale. HRV should be assessed at the broadest scale first, then stepped down to the next level, reassessed, and so on down to the site level.

Another factor complicating the application of HRV is that disturbance itself is multidimensional. Not all patches are created equal. It sounds quite simple to say that the proper distribution of patch ages should be maintained in the managed forest. But the frequency of patch creation is not the only relevant factor. The disturbance regime that creates new patches can be characterized according to type of disturbance, frequency, predictability, size and shape of patches, intensity or severity, and seasonality (Pickett and White 1985; Agee 1993). Each of these factors can have long-lasting effects on the ecosystem. For example, studies in the Pacific Northwest have found that the effects of catastrophic or stand-replacing fires were extremely heterogeneous, creating a diversity of patch sizes, shapes, and configurations, with long-lasting consequences (Morrison and Swanson 1990). Not only does the distribution of patch *ages* need to be maintained within HRV, but so does patch *character*. It is not the creation of even-aged patches through management, per se, that has altered the character of Pacific Northwest forests to their current degraded state, but the frequency, intensity, and dispersion of 16-hectare clear-cut blocks (Hansen et al. 1991). The cumulative effect of changing all these disturbance variables has been the loss of ecosystem integrity. A forest managed within HRV would not only contain significant older forest with the proper configuration, but it would also contain young patches with size, shape, composition, and structure within the HRV of young patches.

HRV is only applicable to management if changes in the factors controlling ecosystems have not permanently altered ecosystem conditions. In some places, fire suppression, grazing, and logging have transformed disturbance regimes. Simply reintroducing disturbance may produce effects outside of any historical precedent, driving the system farther from, rather than closer to, HRV. Likewise, biological invasions have changed the function of whole ecosystems (Vitousek 1986). Our ability to restore historical composition may be seriously constrained by cost, even where we know what we need to do. Finally, the application of HRV may lose all utility under a dramatically altered climate. Where future conditions are not at all like the past, historical behavior is of little use as a management guide.

Summary and Conclusions

Historical range of variability can be a useful tool for thinking about the characteristics of dynamic ecosystems. It can provide a basis for a reserve system and for the management of the intervening matrix if it properly reflects scale dependence and the complexity of the post-disturbance environment. But the application of HRV to conservation is truly a coarse filter. Follow-up fine-filter assessment and conservation measures are absolutely essential to account for biological diversity at all hierarchical levels of organization. The HRV approach may have little utility in the conservation of rare or locally endemic species and assemblages. Used in tandem with a fine-filter approach, HRV provides a scientifically sound rationale for the construction of a comprehensive approach to conservation.

Nevertheless, we must be cautious in developing management prescriptions that ostensibly mimic conditions within HRV. Using the HRV approach means providing a full diversity of structural elements in variable configurations and quantities, with the ultimate objective being maintenance of dynamic patterns and processes that are integral to healthy ecosystems. But we are just beginning to comprehend the complexity of ecosystems; developing prescriptions to sustain that complexity presents a formidable challenge.

Finally, we close with a thought that turns the HRV/coarse-filter relation on its head. That is, not only can HRV be used to develop a coarse filter for land protection, but land protection is also needed as a coarse filter for HRV. Describing HRV requires protection of the remaining areas of high ecological integrity from which we can derive historical conditions. This is not a new concept; Aldo Leopold (1949) described a role for wilderness as "land laboratory" almost fifty years ago. If we are going to use HRV to guide management, protecting those intact parts of the landscape that can provide the key to the past is absolutely imperative. This reason alone is sufficient to demand that reserves be a significant component of any strategy that employs HRV as a guide for management.

Restoring Biodiversity on Public Forest Lands through Disturbance and Patch Management Irrespective of Land-Use Allocation

Richard L. Everett and
John F. Lehmkuhl

Critical elements of biodiversity in heterogeneous forest environments may be distributed in patches over large landscapes. The landscapes themselves are a mosaic of private and public land-use allocations designed to meet numerous public expectations that may or may not be compatible with biodiversity conservation.

Such allocations of land administratively fragment the landscape, making management and conservation efforts complex (Everett et al. 1994; Everett and Lehmkuhl 1996; Landres et al. 1998a). Further, differences in vegetation structure arising from differing standards of use result in ecologically fragmented landscapes (Wiens et al. 1985; Schonewald-Cox and Bayless 1986; Everett and Lehmkuhl 1996; Landres et al. 1998a). While objectives of specific allocations may be met, larger-scale biodiversity objectives may be jeopardized by this fragmentation of habitats (Soulé and Wilcox 1980; Harris 1984), the boundary effects of fragmented habitats (Soulé 1986; Jansen 1986; Wiens et al. 1985; Landres et al. 1998a), and the disruption of disturbance and recovery processes across large landscapes (Schonewald-Cox and Bayless 1986; Everett et al. 1996).

Differing management goals and the resulting ecosystem character of private versus public lands provide great challenges to biodiversity conservation (Noss 1983; Sprugel 1991; DellaSala et al. 1996). However, management of public lands perhaps presents greater challenges because of the wider array of social, economic, and ecological demands placed on them. On public forest lands, the U.S. Forest Service, under the mandate of the National Forest Management Act, developed management plans that allocated land use between a number of competing activities, such as timber production, livestock use, recreation, and wildlife habitat, according to a multiple-use model (Diaz and Apostol 1993; Smith et al. 1995). The law did not specifically define the mechanics of how that allocation should occur on the landscape, so land-use emphasis was defined by permanent allocation for the "best" use. The result is administrative and ecological fragmentation of landscape with similar ecological potential. For example, composition and structure of dry pine-fir forests of the inland western United States that are managed for timber production are different from landscapes managed for deer winter range (40 percent cover, 60 percent forage; Thomas 1979), and both could be significantly different from historical stand conditions that were in synchrony with disturbance regimes (Figure 6.1A, B).

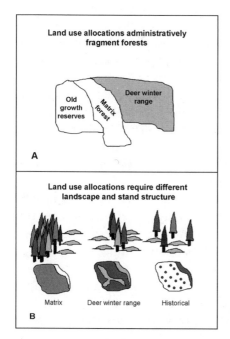

Figure 6.1. A, administrative fragmentation of landscapes by land-use allocations; B, associated stand and landscape characteristics.

Current approaches to federal land-use planning emphasize management of ecosystems for biodiversity conservation and compatible human use (Agee and Johnson 1988; Overbay 1992; Dombeck 1996), but allocations and the problems associated with them remain even where allocations, such as reserves, are primarily intended to preserve biodiversity (Schonewald-Cox and Bayless 1986; Everett and Lehmkuhl 1996; Landres et al. 1998a, b). For example, Camp (1995) found that current amounts of late-successional forest in reserves designated on federal forest land in the eastern Washington Cascades by the Northwest Forest Plan (USDA 1993) were not sustainable over the long term. She also found a similar amount of late-successional forest in adjacent allocations managed under standards that provided even less protection for late-successional forest. The relatively poor prospect of maintaining late-successional forest in the reserves was a function of the highly dynamic disturbance regimes that exist in dry forests of the eastern Washington Cascades (Agee and Edmunds 1992).

Inherent disturbance regimes, defined as the combination of natural disturbance regimes (insects, pathogens, fire, windthrow, mass wasting, indigenous people) and human-induced disturbance regimes subsequent to European settlement, have shaped forest stand and landscape structure and set limits on current management (Everett et al., in press). Goals for allocations of biodiversity conservation, ecosystem integrity, or resource production are unlikely to be achieved when anomalous ecosystem components, such as dry forest habitat for northern spotted owls (Agee and Edmunds 1992), cannot be supported under the inherent disturbance regime. Maintenance or restoration of ecosystem integrity on public lands (Quigley and Bigler Cole 1997) will require the minimum dynamic area (Pickett and Thompson 1978; White 1987) for disturbance regimes to be a primary factor in management strategies, whether those regimes are based on allocations or on the elimination of boundaries to allow a "whole-unit" approach.

Rather than adding more or different allocations and their attendant ecological and administrative problems, we need to expand and integrate biodiversity and social goals across the whole landscape and manage landscape elements in a way that accords each patch the maximum independence of its management protocol and recognizes the ecological limits to resource extraction (Ehrenfeld 1991; Haynes et al. 1996). Under a "whole-unit" ecological approach (Everett and

Lehmkuhl 1996), landscapes are viewed and managed as a unit with a consistent primary objective of ecological integrity for all areas, rather than as an assemblage of land allocations (reserves, matrix forest, riparian, and so forth) with different management standards and resulting variation in integrity. Allocation boundaries are dissolved to the extent ecologically and administratively possible, and pattern and process are managed based on the ecological potential and management opportunities, such as infrastructure, in each patch.

Until a whole-unit approach can be realized, a transitional phase would combine compatible allocations and implement an "emphasis-use" process (Everett et al. 1994; Everett and Lehmkuhl 1996) to integrate land allocations within large landscapes into more functional ecosystems. Under this approach, biodiversity goals are emphasized in similar habitats in adjacent land allocations, and continuity in disturbance regimes is restored among adjacent land allocations. This process promotes the reestablishment of inherent disturbance regimes and the minimum dynamic area required for ecosystem maintenance (Walker 1992). In the long term, the reestablishment and management for inherent disturbance effects lead to "whole-unit management" of the larger landscape.

In the following pages we describe a stepwise process to achieve whole-unit management that begins with integration of existing allocations in a transitional "emphasis-use" approach, then moves to dissolution of allocations for whole-unit management. We describe three steps in the process. Step 1 consolidates adjacent compatible allocations to reduce administrative fragmentation of the landscape. It focuses on similarities in vegetation between adjacent land-use allocations and combines land-use allocations where possible. Step 2 attempts to integrate, but not dissolve, adjacent allocations with significantly different land-use expectations and required vegetation structure. It focuses on matching within-allocation patch heterogeneity with among-allocation patch similarity to expand biodiversity goals and to restore connectivity in disturbance regimes across allocation boundaries. These first two steps work with existing allocations, including reserve systems, and could be implemented immediately.

Step 3 moves to whole-unit management by reinitiating inherent disturbance regimes and managing vegetation pattern and dynamics based on resource potential of the vegetation patch, regardless of current land-use allocation boundaries. The eventual outcome of Step 3 is whole-unit management and the dissolution of allocation boundaries within ecological units.

Step 1: Consolidate Compatible Allocations

The boundaries of land-use allocations often reflect legal mandates and ease of administration rather than precise ecological boundaries and uniformity of vegetation (Keiter 1988; Landres et al. 1998a; Brunson, in press). An allocation captures a percentage of the desired resource condition but may exclude similar areas located in adjacent allocations with different emphasized uses. Moreover, the landscape within the allocation can be heterogeneous, with patches varying in potential, thereby achieving the desired conditions. As a result, an allocation can project unrealistic expectations for the maintenance or creation of desired conditions within its own boundaries and preclude opportunities for using similar conditions in compatible adjacent allocations to meet larger-scale management goals.

Camp et al. (1996) demonstrated the limited ability to grow and maintain late-successional forests within the Swauk Late Successional Forest Reserve in the eastern Washington Cascades, a potential range of 9 to 16 percent of landscape, and showed that a similar limited amount of old forest occurred in similar adjacent, unprotected land allocations (Figure 6.2).

We suggest that the first step in integrated landscape management is to evaluate adjacent land-use allocations for similar current or potential vegetation conditions that meet the emphasized use of each, and then combine allocations where possible. An example would be combining adjacent land allocations for deer winter range and for livestock grazing to improve management on the extensive (57,000 hectares) Tyee Fire in the Entiat Watershed in eastern Washington (Everett et al. 1996; Figure 6.3A, B). Required vegetation for both uses is sufficiently similar that allocations may be combined and both uses achieved from the whole. The scenic allocation could also be integrated with adjacent allocations for increased landscape continuity. The resulting larger management areas should provide greater connectivity in vegetation conditions and common disturbances across the landscape. The capability to manage for unplanned disturbance effects should increase as we increase opportunities to provide resource conditions elsewhere in the undisturbed portions of the larger land base of the combined allocation.

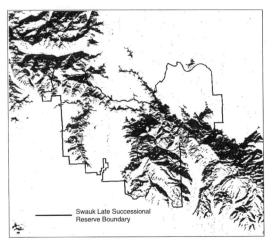

Figure 6.2. Amounts and spatial location of late-successional old-forest patches in a portion of the Swauk Late Successional Forest Reserve (Camp et al. 1996) and surrounding area.

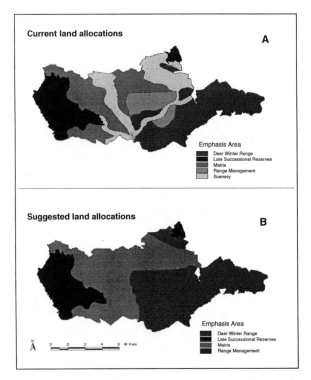

Figure 6.3. Integration of range, deer winter habitat, and scenic allocations on the Tyee burn area in eastern Washington. A, current allocations; B, suggested allocations, combining similar allocations.

Step 2: Integrate Dissimilar Allocations

An "emphasis-use" approach (Everett et al. 1994) to integrating adjacent allocations protects the emphasized use of the individual allocation but also promotes the integration of larger-scale objectives across allocations. The emphasis-use approach integrates adjacent land allocations through reciprocal conservation of emphasized conditions (for example, vegetation patterns, unique habitats) and through the reestablishment of shared disturbance regimes and their effects among allocations so as to increase their sustainability (White 1987) (Figure 6.4). If applied correctly, the approach can potentially reduce administrative fragmentation of forest landscapes by management activities and maintain the dynamic nature of the ecosystems while still allowing harvesting of forest products to meet public expectations (Everett and Lehmkuhl 1996). The strength of the emphasis-use approach lies in its ability to realistically and ecologically manage various types of existing allocations and its potential to integrate diverse allocations (including reserves) to achieve biodiversity goals in the larger landscape. Another advantage of this approach is that it does not require additional allocations and can be implemented immediately.

Extend Biodiversity Goals to Adjacent Allocations

The area required to conserve biodiversity with reserve-type allocations exceeds the amount of pristine land available or that which would likely be dedicated to biodiversity considering other competing land values (Hansen et al. 1991). Given that only 3 percent of the world's surface is in protected areas (Samson 1992), we need to capitalize on opportunities to expand biodiversity objectives to adjacent areas where desired habitats may also be present (DellaSala et al. 1996). The establishment of "late-successional reserves" (LSRs) under the Northwest Forest Plan (USDA 1993) is a partial example of overlaying an old-forest conservation emphasis on an array of existing land-use allocations to extend conservation of old forest beyond existing wilderness and roadless areas. LSRs provide a common conservation emphasis (old forest) among a collection of existing allocations to increase emphasized habitat and reduce reserve isolation. As suggested by Franklin and Forman (1987), the latter process is akin to "feathering the edges" of individual stands. The problem with the

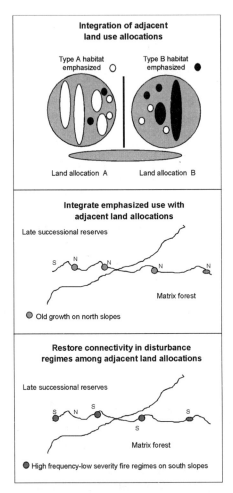

Figure 6.4. Reciprocal conservation of desired habitat across allocations and the restoration of shared disturbance regimes.

LSR solution is that the emphasis occurs in new formal allocations overlaid on old ones and is not applied to the whole landscape, so only sections of existing wilderness or roadless areas are buffered. Consequently, the LSR becomes a new allocation with boundaries and standards and their attendant ecological and administrative problems and complexities.

The emphasis-use process could be repeated at the next lower scale to integrate LSRs with adjacent allocations. For example, feathering the edges of land allocations would benefit late-successional reserves established in dry pine and fir forests on the east slope of the Cascades. Here old-forest stands occur as small patches in specific topographic positions in a matrix of other forest types (see Figure 6.2; Camp 1995; Camp et al. 1996). Because old forests occur on only a small portion of the land surface (9 to 16 percent), the conservation of old-forest habitat in adjacent land allocations would significantly increase amounts of old forest conserved, while leaving the remaining 84 to 91 percent of the adjacent land allocation dedicated to its primary emphasized use. Conserving habitat in adjacent allocations would improve continuity in forest structure, composition, and pattern across the larger landscape. This conservation of habitat in adjacent land allocations could be balanced with management in the LSR allocation to meet the emphasized uses of the adjacent allocation, provided that results are in synchrony with reserve intent, are supported by inherent disturbance regimes, and meet legal mandates.

Protect the Emphasized Use through Disturbance Management

Conserving biodiversity does not mean holding nature static, but rather "perpetuating the dynamic processes of presettlement landscapes" (Noss 1983; Hansen et al. 1991; DellaSala et al. 1996). Disturbance management is directed toward maintaining the balance of disturbance and recovery processes. Whole biotas have evolved under dominant disturbance factors such as fire, and some terrestrial and aquatic systems require a disturbance "pulse" to maintain ecosystem function (Odum 1969). Individual species may require disturbance for survival, as is the case for the rare plant species *Alopecurus aequalis* var. *sonomensis* growing in dynamic dune systems at Point Reyes, California (Fellers and Norris 1991). Disturbance affects ecosystems at many scales, and some, such as fire, work at much larger scales than the area of allocations.

To protect biodiversity, inherent disturbances both internal and external to the target allocation need to be restored and managed to reduce deleterious effects. Deleterious effects—the loss of structure, function, or species—can arise from both excessive and insufficient disturbance. Under excessive disturbance levels, disturbance reoccurs before the restoration process is complete and site degradation occurs (Turner et al. 1993). Too frequent or severe prescribed burning or timber harvest can deplete site nutrient capital, cause soil compaction, or reduce coarse woody debris and lower site potential for biomass production (Harvey et al. 1981; Grier et al. 1989). On a landscape basis, excessive disturbance of old-forest habitats has reduced amounts below historical levels (Everett et al. 1994).

Insufficient disturbance occurs when disturbance characteristic of the biophysical environment is suppressed. An example is fire suppression on the east slope of the Washington Cascades that resulted in stand development beyond that supported by the inherent disturbance regimes (Agee and Edmunds 1992; Everett et al., in press). When the system eventually corrects itself, the effects may be more severe and of greater extent than what historically occurred, with the potential for catastrophic loss in habitats (Covington et al. 1994; Everett et al. 1996) and site nutrient capital (Grier 1975).

Disturbance management needs to consider the different spatial scales and intensities of coexisting disturbance regimes. Disturbance management to conserve the limited and patchy (less than 1 hectare) habitat of Wenatchee larkspur (*Delphinium veridescens*, threatened species) in the Camas Meadows of eastern Washington is an example of managing for hierarchical disturbance effects at different spatial scales and levels of intensity. At the site level, within-patch disturbance is required to prevent tree canopy closure and loss of habitat, but disturbance needs to be moderate to maintain canopy shading at between 33 and 66 percent for maximum *Delphinium* vigor (Kuhlmann and Everett, in press). At the watershed level, excessive or insufficient disturbance that causes a significant loss or gain in tree cover could alter hydrologic processes and adversely affect *Delphinium* habitat, which is restricted to moist soils adjacent to streams, seeps, and shallow drainage bottoms. At still larger drainage basin scales, the grazing and trampling effects of an expanding elk herd that utilizes several watersheds need to be considered. Although the area of *Delphinium* habitat may be small, "the area of concern" where disturbance is managed for species viability is much larger (Figure 6.5).

Increase Disturbance Continuity

Reserves will probably remain too small to contain all required diversity (Noss and Cooperrider 1994), and it is unlikely they will capture the "minimum dynamic area" of some of the forms of disturbance required for long-term reserve maintenance (Pickett and Thompson 1978; White 1987). This suggests the need for disturbance management at larger scales than the reserve, or an "expanded coarse filter" (Noss 1987; DellaSala et al. 1996), to conserve biodiversity outside

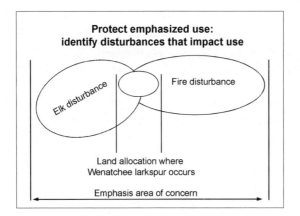

Figure 6.5. Hierarchical disturbance effects on habitat of the rare Wenatchee larkspur (*Delphinium veridescens*).

reserve boundaries. Rather than enlarging the reserve boundaries to encompass large-scale disturbance events, only the disturbance boundary (area of concern; Figure 6.5) need be expanded into the adjacent allocations. This disturbance boundary can be viewed as one of the multiple boundaries to conserve biodiversity within and outside reserve areas.

For those disturbances whose historical extent affected the area of several land allocations, a policy to enhance continuity in disturbance effects across boundaries (e.g., broad "let burn" policy) is desirable to maintain landscape integrity. In other instances, historical disturbance was patchy in nature (e.g., root rot; Hessburg et al. 1994), and individual events were confined within each allocation. Here, maintaining the patchy disturbance process within each allocation would preserve the larger landscape-level process and effects. In the LSR example, continuity in disturbance regimes could be accomplished by the reestablishment of high-frequency/low-severity fire regimes on southern slopes in both the LSR and adjacent allocations (Figure 6.4). Periodic ground fires could maintain low crown fire hazard in the reserve and adjacent allocations such that neither presents a fire hazard to the other.

Precautions

The emphasis-use concept fails when the initial land allocation is faulty—that is, when the allocation and the emphasized use are not in synchrony with inherent disturbance regimes. For example, if late-successional reserves are established in areas with a high-frequency/low-severity fire regime, we should not expect to be successful in creating and maintaining the dense, multilayered old-forest structure and composition that would occur only under a low-frequency/high-severity fire regime.

The reevaluation of land allocation, emphasized use, and the sustainability of required vegetation structure under inherent disturbance regimes of the area should be a priority for all land allocations and specifically for biodiversity reserves. If reserves that have been created are not sustainable over time, new reserve areas should be found or created to safeguard biodiversity over the long term. Reserve designations based on current vegetation characteristics may not have considered that existing vegetation is a result of altered disturbance regimes and that "spurious habitat" may not be supportable under the inherent disturbance regimes of the area over the long term (Agee and Edmunds, 1992; Everett et al. 1997). Utilization of spurious habitat may be required in the short term to maintain viability of species dependent upon habitat that has been lost elsewhere, but we suggest that other, more stable reserve areas be identified to maintain species over time.

The reevaluation of reserve areas and their modification were an integral part of the Northwest Forest Plan (USDA 1993) east of the Cascades and also of the PACFISH (USDA and USDI 1994) buffer zones for riparian areas. In both instances, the creators of these documents realized the dynamic nature of inland forests and recommended the reevaluation of initial set-asides as more scientific information became available. New information on inherent disturbance regimes and vegetation characteristics that are in synchrony with the accompanying disturbance effects will improve the land manager's ability to evaluate the need, characteristics, and sustainability of reserve or buffer areas over time.

Step 3: Move to Whole-Unit Management

> Managing regional (and smaller) landscapes as a whole is preferred when trying to further the goal of biodiversity while meeting the human need for natural resources. (DellaSala et al. 1996)

"Whole-unit management" is defined as management to achieve the array of public expectations for resource conditions and products from the whole ecological management unit rather than compartmentalizing desired resource conditions and products among the individual land-use allocations. According to Stoszek (1990), "Forest stands within the unit would be treated as parts of one interacting entity, not as isolated planning segments of conventional forestry." The goal would be to manage for fully functional and interconnected ecosystems (Noss and Cooperrider 1994) free of the ecological and administrative problems associated with boundaries.

The whole-unit concept is based on the assumption that resource conditions in dynamic forest systems are rarely consolidated in finite areas over extended lengths of time but are continually redistributed throughout the landscape in accordance with topographic, edaphic, biotic, and disturbance influences. Managing the disturbance regime is key to whole-unit management.

Disturbance Management

Reestablishment of natural disturbance processes should maintain or enhance the dynamic nature, integrity, and long-term stability of ecosystems and the conservation of biodiversity (Leopold et al. 1963; Noss 1983; Parsons et al. 1986; Urban et al. 1987; Agee and Johnson 1988; Christensen 1988; Samson 1992). Agee and Huff (1985) suggested the shepherding of natural disturbance patterns across the landscape for intelligent management of wilderness. However, realistic expectations and goals are needed in defining our abilities to conduct disturbance management and the reinitiating of inherent disturbance regimes.

Hunter (1987) recommended the use of natural disturbance as a relevant model for managing spatial heterogeneity, but in the application of large-crown fire disturbance regimes, he pointed out the need for practicality because of current social constraints (Hunter 1993). In nonequilibrium landscapes, large-scale loss of reserves or other allocations for extended periods of time would not meet public expectations. Also, the ability to reestablish large-scale disturbance regimes may no longer be possible with altered landscapes, and so the reestablishment of large-scale disturbance may have to be created through the cumulative treatment of smaller areas (Swanson and Franklin 1992). Reestablishing inherent disturbance regimes in ecosystems characterized by a dynamic equilibrium in shifting mosaic vegetation would require mimicking the frequency, severity, and extent of the dominant disturbance agents. Moreover, a great deal of flexibility would be required in disturbance management because of unplanned disturbance events in dynamic inland forest systems. However, we would be managing for change, a rare constant in natural systems (Agee and Johnson 1988), rather than nonviable, static systems.

We are rapidly increasing our ability to define disturbance regimes both in the present and past. Disturbance regimes can be identified and quantified through the study of vegetation patterns (Swanson et al. 1994), stand composition, age structure and gap dynamics (Lundquist 1995; Spies et al. 1990), and tree ring analysis for fire and insect events (Fritts and Swetnam 1989). Composite disturbance regimes are now being formulated for some areas (Wargo 1995), and disturbance profiles are used to define causal factors and results in stand dynamics (Lundquist 1995). Disturbance response surfaces are being developed to describe the probability of disturbance events based on their frequency, severity, and extent (Swanson et al. 1994; Everett et al., in press). Information on historical vegetation patch composition, size, and spatial location would provide insight into historical disturbance effects and regimes (Morgan et al. 1994) and define when disturbance and resulting patch conditions are significantly over- or underrepresented on the landscape (Haufler et al. 1996). This information could be used to define where disturbance effects are required to create desired habitat and increase disturbance continuity to maximize biodiversity goals and other public expectations.

Social and Economic Acceptance of Whole-Unit Management

Land use has been described as the intersection of three planes: ideological, social, and physical (activities) (Smith et al. 1995). The increase in land-use allocations in response to continually emerging public issues graphically demonstrates the complexity of public expectations for land use. In this context, whole-unit management is an ideological compromise between biocentric and anthropocentric resource management philosophies. Nature is allowed to take its course under the inherent disturbance regimes of the area, but the rate and direction of change providing for the maintenance of ecosystem integrity are used to encourage desired resource conditions and forest products. The process appears socially acceptable because it provides for desired human habitats and utilizes the resource base to meet multiple socioeconomic expectations. It also advances the goal that our physical effects on the landscape reflect management working with natural processes of disturbance and recovery and provide for long-term maintenance and integrity of the ecosystem. Socioeconomic gains are maximized when management activities are consistent with the functioning of ecosystems (Averill et al. 1997).

Implementing Whole-Unit Management

Whole-unit ecosystem management approaches are not new. For millennia, the Yakama Indian Nation of eastern Washington practiced an integrated land-use approach that recognized different resource potentials among vegetation patches and emphasized uses within large landscapes but allowed each piece to contribute to their social, cultural, and economic well-being (Uebelacker 1986). Their extensive knowledge of the resource base, its spatial–temporal relationships, and the freedom they had to use centers of resource convergence made this possible. Such coarse-filter approaches of conservation (Hunter 1987; Noss and Cooperrider 1994; Haufler et al. 1996) are also forms of whole-unit management.

Implementation of whole-unit types of management that deemphasizes allocations might resemble the work of the Interior Columbia Basin Ecosystem Project (Quigley et al. 1996). On this project, spatially referenced information on resource conditions and disturbance regimes were assembled and analyzed to allow land-use integration across multiple administrative and jurisdictional boundaries to achieve large-scale social, economic, and resource-based objectives. Multidisciplinary information was gathered to assess conditions and develop management alternatives at multiple scales: region, ecoregion, drainage basin, and individual watershed. A multiscale analysis of the interaction of the biophysical environment, ecological processes (particularly disturbance), past history, and management needs and opportunities was used to target management approaches on a watershed level. Under the "active management" alternative, areas with moderate-to-high ecological integrity were given a conservation emphasis. Those areas would act as sources for ecosystem services, processes, and wildlife populations. The conservation emphasis would prescribe relatively minimal management, similar to reserves. However, unlike many reserve proposals (Noss 1993; Noss and Cooperrider 1994), management standards and

boundaries would be flexible to account for unpredictable disturbance events that might alter the value of the area for conservation
(for example, wildfire in old-forest reserves) and for needed restoration within the area. Other areas with greater human disturbance
would be managed through some level of active restoration based on
their biophysical potential, need for restoration, and the opportunity
for management, such as road systems and other infrastructure.
Production of goods and services on federal lands would be an outcome of restoration activities, but production might remain a primary emphasis on private lands. Regardless of emphasis (conservation,
restoration, production), the goal of ecosystem integrity would be
the same and emphasis areas would share common goals—for example, maintenance and restoration of old forest, riparian systems, and
disturbance regimes. Analysis of such an active approach (Haynes et
al. 1996) showed that ecological integrity (Quigley et al. 1996) and
vertebrate species viability (Lehmkuhl et al. 1997) would be similar
to that of a large regional reserve network with relatively passive management.

Whole-unit management is currently being developed for 16,000
hectares of the Teanaway Project area on lands owned by the Boise
Cascade Corporation in eastern Washington and in Idaho (Haufler et
al. 1996). All stands have been typed and mapped to describe ecosystem characteristics and potential forest products, resource conditions,
and species viability. Both this and a related project in the Idaho
Southern Batholith are using an ecosystem diversity matrix of vegetation patches and knowledge of inherent disturbance regimes to
define conditions needed for long-term maintenance of their forest
systems. The ecosystem matrix provides information on what constitutes the adequate ecological representation of vegetation development stages on the landscape and how it can be used to integrate the
company's landholdings with adjacent lands and define its role in biodiversity conservation within the larger ecological unit.

Summary and Conclusions

A practical and flexible approach for managing wildlands and natural
resources will be found in the middle ground of ecosystem management that attempts to integrate diverse human values and uses while
maintaining or restoring ecological integrity. A reserve model with
core, buffer, and multiple-use areas may be flexible enough to inte-

grate biological and social goals. However, boundaries have ecological and management costs, and critical elements of biodiversity may be dispersed over large landscapes and may not occur or be maintained in existing reserves. Larger and more numerous reserves might enhance conservation of biodiversity, but this option may hinder our ability to meet other ecosystem objectives. Moreover, more or larger reserves may not be a practical or ecologically sustainable solution in dynamic landscapes or those in which humans have already highly altered ecological pattern and processes. Reserve network management, which should be primarily very conservative, needs also to be flexible to restore areas of past human abuse within reserves, and areas outside reserves need to be managed to account for patterns and processes working at scales larger than individual reserves. The reserve model becomes fuzzy when boundaries need to be porous to disturbance. In point of fact, any practice is acceptable if it maintains or improves the ecological integrity of an area.

We suggest viewing the landscape as a whole unit with a consistent primary objective of ecological integrity for all areas, rather than as an assemblage of allocations (core areas, buffers, and so forth) with different levels of management and resulting integrity. Conservation or restoration of ecosystem processes, primarily inherent disturbance regimes, would be a principal objective. Problems associated with the stability of reserves in dynamic landscapes would be minimized, and conservation would not be constrained by continuing processes of disturbance and renewal that have historically maintained ecosystem integrity. As part of this strategy, reserve-like areas with a conservation emphasis might be established in areas of moderate-to-high integrity where natural processes dominate, where conservation is the primary goal, where resource extraction activities have not yet significantly altered ecosystems but would have a significant effect on them, and where the risk of losing the reserve to disturbance is small. Reserves would be a tool, not a primary objective.

Others have proposed similar models that attempt to manage ecosystems by relying less on delineating permanent land allocations with standard management prescriptions than on the potential of landscape elements to integrate ecological and social goals across the landscape. This "whole-unit" approach is not a disguised multiple-use paradigm but is a true ecosystem approach that puts ecological integrity first and recognizes the ecological limits to resource extraction. It stresses whole-landscape management by prescribing passive (conservation reserves), active, or production (commodity) management strategies based on a multiscale analysis of the interaction of the biophysical environment, ecological processes (particularly disturbance), past history, and management needs and opportunities. Areas with moderate-to-high integrity likely would have a conservation emphasis similar to a core reserve and act as source areas for ecosystem services and processes. However, the conservation emphasis would be flexible, rather than static, to account for unpredictable disturbance events. Other areas with greater human disturbance would be managed according to their potential and need for restoration. Production of goods and services on federal lands would be an outcome of restoration activities, but production might remain a primary emphasis on private lands.

There are three phases to initiating whole-unit management:

1. Consolidate adjacent compatible land-use allocations and reduce administrative fragmentation of the landscape.

2. Integrate adjacent land-use allocations with significantly different land-use expectations and required vegetation structure.

3. Move to whole-unit management by dissolving land-use allocation boundaries within ecological units, reinitiating inherent disturbance effects, and managing vegetation pattern and dynamics based on resource potential of the patch.

Step 1 identifies similarities in vegetation in order to meet the emphasized use of adjacent land-use allocations and combines land-use areas for ease of disturbance management. Step 2 focuses on matching within-allocation patch heterogeneity with among-allocation patch similarity to expand biodiversity goals and to restore connectivity in disturbance regimes across allocation boundaries. These first two steps work with existing allocations, or reserve systems, and could be implemented immediately to move toward whole-unit management. Step 3 focuses on the potential of individual patches to contribute to biodiversity or other resource conditions regardless of the land-use allocation in which they currently occur.

Recent efforts to craft a conservation strategy for federal lands in the interior Columbia River Basin considered both the conventional reserve and whole-unit approaches. Evaluation of the alternatives found a whole-unit approach to be slightly better overall in maintaining or restoring critical elements of biodiversity while providing more value for human use than a large regional reserve system.

To summarize, this chapter describes a stepwise process that integrates competing current uses and disturbance regimes across large landscapes and achieves ecological integrity and public expectations on a whole ecological unit basis (that is, watershed, drainage basin).

Chapter 7

Conserving Biological Diversity Using a Coarse-Filter Approach with a Species Assessment

Jonathan B. Haufler,
Carolyn A. Mehl, and
Gary J. Roloff

Various strategies have been proposed to address the conservation of biological diversity. Each strategy has different assumptions (Haufler, Chapter 2), and each has various advantages and disadvantages (Haufler, Chapter 14). But strategies for conserving biological diversity will be of questionable effectiveness in practical implementation if they fail to incorporate human needs for commodities or social desires. For this reason, strategies for conserving biological diversity are often incorporated into ecosystem management initiatives.

Ecosystem management generally strives to provide an optimal mix of ecological, social, and economic objectives for a given landscape (Kaufmann et al. 1994). Maintaining or restoring regional biodiversity at genetic, species, and community levels is a critical component for meeting the ecological objectives of ecosystem management. This chapter discusses a process for ecosystem management that provides for the maintenance or restoration of biological diversity while also allowing social and economic objectives to be incorporated into the process. This strategy utilizes a coarse filter applied at a landscape scale and provides a mechanism to check the coarse filter through individual species assessments. The coarse-filter component emphasizes ecological communities, whereas the species assessment addresses species-specific habitat needs for population viability. We present

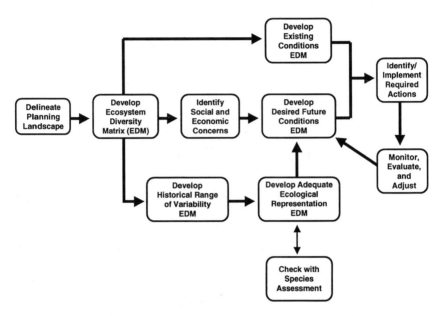

Figure 7.1. A process for ecosystem management designed to address the conservation of biological diversity as well as incorporate economic and social objectives. The process uses a coarse-filter approach with a species assessment as a check.

this strategy with primary reference to the forested communities of west-central Idaho, where Boise Cascade Corporation is using this process in a collaborative ecosystem management project.

An Ecosystem Management Process

The basic process described here (Figure 7.1) was previously discussed by Haufler et al. (1996). Ecological objectives of biodiversity conservation are addressed by means of a coarse-filter approach linked with a species assessment, the process of which is outlined as follows:

1. Delineate the planning landscape.
2. Develop the ecosystem diversity matrix (EDM).
3. Develop historical range of variability EDM.
4. Develop adequate ecological representation EDM.

5. Check adequate ecological representation EDM with species assessments.

6. Identify social and economic concerns.

7. Develop desired future condition EDM.

8. Develop existing condition EDM.

9. Identify and implement required actions.

10. Monitor, evaluate, and adjust through adaptive management.

Delineate the Planning Landscape

Our use of the term "planning landscape" follows Haufler's definition. Identifying and delineating an appropriate-sized planning landscape should be based primarily on ecological objectives. Social and economic objectives operate at a variety of levels often independent of ecological boundaries, requiring consideration of multiple factors at multiple spatial scales (Haufler et al., in press). For instance, demands for commodities may be influenced by local, national, and global economies and markets. The demands for social and economic uses of public lands by local communities are often different from and may be overridden by national demands for other uses.

Ecological objectives for maintaining and enhancing biological diversity also function at multiple scales. For instance, maintaining a viable population of wolves (*Canis lupus*) will require a large area compared to that needed for a viable population of snowshoe hare (*Lepus americanus*). Because of issues of scale like these, identifying a hierarchical structure for landscape delineation is important. The National Hierarchy of Ecological Units (Ecomap 1993) provides one example of a hierarchical ecological classification.

While multiple spatial scales must be considered in meeting ecological objectives, one must still delineate a primary planning landscape in order to implement a coarse-filter strategy. It is also of equal importance to identify the criteria for delineation. Haufler et al. (1996) advocated the delineation of a planning landscape based on ecological boundaries that balanced four primary considerations:

- Similar biogeoclimatic conditions that influence site potentials.

- Similar historical disturbance regimes that influence vegetation structures and species compositions.

- Adequately sized landscape providing a sufficient range of habitat conditions to assure population maintenance of the majority of native species that historically occurred in the planning landscape, excluding certain species such as megafauna. Megafauna or species with very low population densities will require analyses at broader scales, where contributions from landscapes are aggregated to address population maintenance of those species.

- Recognition of maximum size to avoid practical operational limitations in terms of data management, implementation restrictions, and number of cooperating landowners necessary for successful planning.

The fourth consideration has an additional practical component. The planning landscape cannot become so large and the delineation of meaningful, homogeneous ecological units so complex that landscape assessment and management become too difficult for practical implementation. In other words, if the planning landscape becomes very large, it will include a large amount of ecological variability caused by differences in geological or climatic factors, resulting in a logistically unmanageable number of ecological units in the planning landscape. Conversely, if a larger planning landscape were delineated and the number of ecological units restricted to provide management feasibility, then the degree of homogeneity within an ecological unit would suffer. As unit descriptions become more generalized and lumped together, the variance around ecological community or habitat attributes becomes increasingly large, to a point where ecological communities lose their effectiveness in defining species distributions or habitat characteristics. If this occurs, the effectiveness of the coarse filter is compromised.

Ecological units, the building blocks of any coarse-filter approach, are influenced on a broad scale by geologic and climatic factors encompassed in the planning landscape. Each ecological unit is further influenced by site-specific biotic and abiotic factors that ultimately influence the ecological characteristics of the unit. In this way, planning landscapes define the range of specific ecological units that can occur.

Watersheds have been proposed by many as appropriate planning landscapes. Watersheds are certainly important areas to delineate as they provide the network for connectivity of aquatic communities or certain metapopulations. However, as a basis for defining a planning landscape, watersheds have a major drawback. The only unifying feature of watersheds is that the water in the area flows out a common outlet. The watershed, depending upon its size and location, may cross a number of significant ecological boundaries such as geologic formations, and these boundaries can produce dramatic differences in ecological communities. For example, many watersheds that originate in the Idaho Southern Batholith, an area characterized by eroding granitics, ultimately flow into geological areas characterized by basalts. The resulting ecological communities, while still in the same watersheds, can be substantially different, making the delineation of meaningful homogeneous units more complex at the landscape scale. Consequently, to meet ecological objectives, planning landscapes should be delineated on ecological boundaries that help to define the composition of ecological units, rather than by topographic features alone.

Ecomap (1993) provides a hierarchical basis for delineating ecological boundaries for planning landscapes. We have found that the section level, or aggregates of subsections as directed by Ecomap, represents the best ecological boundaries to balance the criteria discussed above. In Idaho, the Idaho Southern Batholith planning landscape, which is an aggregate of several subsections within the Idaho Southern Batholith Section (McNab and Avers 1994), was selected as the planning landscape for a cooperative ecosystem management project (Haufler et al. 1996). This area, approximately 2.4 million hectares, is expected to be large enough to support viable populations of nearly all the native species of the area, with the exception of a few large carnivores. Yet this planning landscape is not expected to be so large as to logistically impede cooperative efforts at ecosystem management.

Develop the Ecosystem Diversity Matrix (EDM)

A coarse-filter approach to conserving biodiversity attempts to provide a mix of natural ecological communities across a planning landscape that will sustain the viability of native species dependent on those systems. A well-developed coarse filter should have several characteristics. It should quantify the natural ecological communities in a landscape such that by providing the different communities, maintenance of biodiversity is ensured. Further, effective coarse filters should not allow subdivision of unique ecological communities that might exclude species. For instance, if old growth were a category in a coarse filter without finer delineation, the filter might only function effectively if old growth were consistently and systematically provided across a landscape. However, if old growth were provided based on elevation breaks, so that all old growth were placed in high elevations, then species dependent upon low-elevation old-growth communities would be excluded and the coarse filter would fail. Conversely, a coarse filter should not be so detailed as to prescribe redundant components. In other words, if a coarse filter described two different structural stages, one dominated by 20- to 30-centimeter trees and the other by 30- to 40-centimeter trees, but no differences in species associations were found between these two stages, they would not need to be distinguished from each other in the filter.

The purpose of developing the EDM is to identify and characterize all the ecological communities occurring under the historical disturbance regimes within the planning landscape. This characterization will be based on the landscape classification system selected. While there are numerous classification systems, land classification systems generally fall into two categories: classifications of existing vegetation, and classifications of ecological parameters defining site conditions. Classifications of existing conditions can be based on structural characteristics (for example, see O'Hara et al. 1996) or on species composition of the dominant vegetation (see, for example, Daubenmire 1952). Structural classifications often incorporate temporal or successional factors, allowing dynamics of vegetation growth stages to be incorporated into the classification. Classifications based on dominant species compositions allow for existing plant communities to be delineated. Both of these types of classifications allow for the mapping of existing vegetation into groupings of varying homogeneity.

In contrast, classification systems based on ecological parameters (ecological classification) allow landscapes to be stratified into different site conditions that then influence the plant and animal communities that will occur on a site. These conditions (for example, soil moisture, slope, aspect) may also influence the characteristics of different disturbance regimes across the landscape. "Habitat typing" (Daubenmire 1968; Steele et al. 1981) is an ecological classification that identifies indicators of areas that will support similar potential vegetation. Habitat types are usually named according to the late-successional vegetation that would be expected to occur at a site. The National Hierarchy of Ecological Units (Ecomap 1993) delineates sites having similar ecological characteristics based on geologic, climatic, soil, and water influences. This classification is hierarchical in that finer-resolution units are delineated within larger units based on differing ecological boundaries and parameters. Historical disturbances can also be characterized at certain levels of the hierarchy. A limitation of both habitat typing and the National Hierarchy of Ecological Units is that neither classification identifies existing vegetation or incorporates temporal dynamics, both important factors for conserving biodiversity. Thus, both existing vegetation classifications and ecological site classifications have advantages and limitations.

Haufler (1994) described a tool known as the "ecosystem diversity matrix" (EDM). This matrix combines both a classification of existing vegetation based on structural characteristics and species composition as well as an ecological classification incorporating biogeoclimatic conditions. In addition to these classifications, the EDM is used to describe past disturbance regimes and the ecological communities produced by these disturbance regimes (Haufler 1994; Haufler et al. 1996). The EDM also includes characterization of ecological communities that might not have occurred under historical disturbance regimes but resulted from successional changes caused by suppression of historical disturbances and that are present in existing conditions. An example of an EDM for forested systems of the Idaho Southern Batholith planning landscape is provided in Figure 7.2. For a more complete description of the Idaho Southern Batholith EDM for forested systems, refer to Haufler et al. (1996).

To provide a complete ecological characterization of the planning landscape, we envision the use of four integrated EDMs. An example of the EDM for forested ecosystems is illustrated in Figure 7.2; we have also developed a matrix of riparian and wetland communities as well as the framework for an aquatic community matrix. In addition, for the Idaho Southern Batholith, a matrix that characterizes drier ecosystems (grass and shrubland habitat types) must also be developed. These four matrices should provide a coarse filter for all ecological communities occurring in the Idaho Southern Batholith, with

VEGETATION GROWTH STAGES	Habitat Type Class: Dry Ponderosa Pine / Xeric Douglas fir — Potential Cover Types	Acres	Warm, Dry Douglas fir / Moist Ponderosa Pine — Potential Cover Types	Acres	Cool, Dry Douglas-Fir — Potential Cover Types	Acres	Dry Grand Fir — Potential Cover Types	Acres	Warm, Dry Subalpine Fir — Potential Cover Types	Acres	Warm, Moist Subalpine Fir — Potential Cover Types	Acres	High Elevation Subalpine Fir — Potential Cover Types	Acres
Grass/Forb/Seedling														
Shrub/ Seedling (L, M, H)	*Pinus ponderosa*		*Pinus ponderosa* (*Pinus contorta*) (*Populus tremuloides*)		(*Pinus contorta*) (*Populus tremuloides*)		*Pinus ponderosa* *Pseudotsuga menziesii* (*Populus tremuloides*)		*Pseudotsuga menziesii* (*Pinus contorta*) (*Populus tremuloides*)		(*Pinus contorta*) (*Larix occidentalis*)		*Pinus albicaulis*	
Sapling; shrub/seedling (L, M, H)	*Pinus ponderosa*		*Pinus ponderosa* (*Pinus contorta*) (*Populus tremuloides*)		(*Pinus contorta*) (*Populus tremuloides*)		*Pinus ponderosa* *Pseudotsuga menziesii* (*Populus tremuloides*)		*Pseudotsuga menziesii* (*Pinus contorta*) (*Populus tremuloides*)		(*Pinus contorta*) (*Larix occidentalis*)		*Pinus albicaulis*	
UNDERSTORY FIRE REGIME	*Understory Burn* *5 -25 Years*		*Understory Burn* *10-22 Years*		Understory Burn 25-100 Years		*Understory Burn* *10-30 Years*		*Fire Mosaic* *50-90Years*		*Some* *Understory Burn*		*Understory Burn* *25-70 Years*	
Small Trees (L, M, H)	*Pinus ponderosa*		*Pinus ponderosa* (*Pinus contorta*) (*Populus tremuloides*)		(*Pinus contorta*) (*Populus tremuloides*)		*Pinus ponderosa*		*Pseudotsuga menziesii* (*Pinus contorta*) (*Populus tremuloides*)		(*Pinus contorta*) (*Larix occidentalis*)		*Pinus albicaulis*	
Medium Trees (L, M, H)	*Pinus ponderosa*		*Pinus ponderosa* (*Pinus contorta*)		*Pseudotsuga menziesii* (*Pinus contorta*)		*Pinus ponderosa*		*Pseudotsuga menziesii* (*Pinus contorta*)		(*Pinus contorta*) (*Larix occidentalis*)		*Pinus albicaulis*	
Large Trees (L, M, H)	*Pinus ponderosa*		*Pinus ponderosa*		*Pseudotsuga menziesii*		*Pinus ponderosa*		(*Pinus contorta*)		(*Larix occidentalis*)		*Pinus albicaulis*	
STAND DESTROYING FIRE REGIME	*Stand Destroying* *Wildfire Unlikely*		*Stand Destroying* *Wildfire Unlikely*		*Some Stand* *Destroying Wildfire*		*Stand Destroying* *Wildfire Unlikely*		*Fire Mosaic* *50-90 Years*		*Stand Destroying* *Wildfire*		*Stand Destroying* *Wildfire*	
Small Trees (L, M, H)	*Pinus ponderosa* (*Pseudotsuga menziesii*)		*Pinus ponderosa* (*Pinus contorta*) (*Populus tremuloides*) (*Pseudotsuga menziesii*)		*Pseudotsuga menziesii* (*Pinus contorta*) (*Populus tremuloides*)		*Pinus ponderosa* *Pseudotsuga menziesii* *Abies grandis* (*Populus tremuloides*)		*Pseudotsuga menziesii* (*Pinus contorta*) (*Populus tremuloides*)		*Pseudotsuga menziesii* *Picea engelmannii* (*Larix occidentalis*) (*Pinus contorta*)		*Pinus albicaulis*	
Medium Trees (L, M, H)	*Pinus ponderosa* (*Pseudotsuga menziesii*)		*Pinus ponderosa* *Pseudotsuga menziesii*		*Pseudotsuga menziesii* (*Pinus contorta*)		*Pinus ponderosa* *Pseudotsuga menziesii* *Abies grandis*		*Pseudotsuga menziesii* *Picea engelmannii* (*Pinus contorta*)		*Pseudotsuga menziesii* *Picea engelmannii*		*Abies lasiocarpa* *Pinus albicaulis* *Picea engelmannii*	
Large Trees (L, M, H)	*Pinus ponderosa* (*Pseudotsuga menziesii*)		*Pinus ponderosa* *Pseudotsuga menziesii*		*Pseudotsuga menziesii*		*Pinus ponderosa* *Pseudotsuga menziesii* *Abies grandis*		*Picea engelmannii* *Abies lasiocarpa*		*Pinus ponderosa* *Pseudotsuga menziesii* *Abies grandis*		*Abies lasiocarpa* *Pinus albicaulis* *Picea engelmannii*	
Old Growth (L, M, H)	*Pinus ponderosa* (*Pseudotsuga menziesii*) ≥ 5 trees ≥ 24" dbh/ac. Snags - infrequent		*Pinus ponderosa* *Pseudotsuga menziesii* ≥ 10 trees > 24" dbh/ac ≤ 1 Snag ≥ 20" dbh/ac		*Pseudotsuga menziesii* ≥10 trees ≥ 18" dbh/ac ≤ 3 Snags ≥ 16" dbh/ac		*Abies grandis* ≥ 15 trees ≥ 24" dbh/ac ≥ 2 Snags ≥ 20" dbh/ac		*Abies lasiocarpa* *Picea engelmannii* >25 trees >24" dbh/ac ≥ 2 Snags ≥ 12" dbh/ac		*Abies lasiocarpa* *Picea engelmannii* >25 trees >24" dbh/ac ≥ 2 Snags ≥ 12" dbh/ac		*Abies lasiocarpa* *Pinus albicaulis* *Picea engelmannii* ≥ 10 trees ≥ 12" dbh/ac	

Figure 7.2. A simplified example of the forested ecosystem diversity matrix for the Idaho Southern Batholith landscape. Old-growth definitions are based on Hamilton (1993). Seral species potentially occurring in an ecological land unit (cells in the matrix) are represented by parentheses; late-successional species have no parentheses. Canopy coverage is represented by L = low, M = medium, and H = high.

the exception of very rare ecological communities (such as cave and talus communities).

Develop Historical Range of Variability EDM

A step that is often overlooked in some strategies for conserving biodiversity is an analysis of the historical range of variability for the planning landscape. This analysis should assist in understanding past disturbance regimes and their influence on the composition, structure, and spatial distribution of vegetation. Within any given landscape, several different disturbance regimes may have operated to influence ecological communities and their spatial distribution. Examples of common disturbance mechanisms in North America include fire, hurricanes, insects, blowdowns, and flooding. The frequency with which disturbance occurs, as influenced by climatic or physical factors, may result in different regimes operating within a given landscape.

As described in the previous section on developing an ecosystem diversity matrix, historical disturbance regimes must be identified within the ecological classification system. In the EDM for forested systems of the Idaho Southern Batholith landscape (Figure 7.2), fire is the primary disturbance mechanism affecting ecological communities across the planning landscape. The biogeoclimatic site conditions that influence vegetation characteristics have also influenced the characteristic intensity and frequency of fire within a habitat-type class. Consequently, the EDM describes ecological units in terms of vegetation characterized by historical disturbance regimes. The intensity of fire and frequency with which it occurred across the Idaho Southern Batholith is represented by a low-intensity, high-frequency fire regime (that is, an understory fire regime) and a high-intensity, low-frequency (that is, a stand-replacing) fire regime.

The value of historical range of variability in maintaining or restoring biodiversity cannot be overemphasized. As discussed in Haufler et al. (1996), various methods and challenges exist for determining the historical range of variability. A historical range of variability matrix can be developed by means of an EDM. With historical disturbance regimes characterized, each ecological unit should be evaluated for its historical range of variability. This consists of determining the percentage range of the area in a particular habitat-type class that occurred within each vegetation growth stage over a designated historical time interval (Haufler et al. 1996).

The development of a historical range of variability EDM is critical for understanding changes that have occurred in the landscape from conditions influenced by historical disturbance regimes. The EDM framework will provide sufficient detail to capture the range of ecological communities to which native species evolved and adapted. Mapping habitat-type classes in a GIS (geographic information system) will also allow an analysis of the spatial distribution of expected historical ecological communities across the planning landscape.

Develop the Adequate Ecological Representation EDM

"Adequate ecological representation" is defined as the sufficient size and distribution of inherent ecosystems to maintain viable populations of all native species dependent on those ecosystems (Haufler 1994). Determination of adequate ecological representation is based on an understanding of historical disturbance regimes and the characteristics of the ecological units that occurred under those regimes (Haufler et al. 1996). Adequate ecological representation is quantified as required acreage, appropriately distributed within the landscape, for any cell in the matrix. For most landscapes, adequate ecological representation can be determined as a fixed percentage of the historical range of variability. A threshold level of the landscape provided under historical ranges of variability should be identified and will serve as a standard level to check against when selected species assessments are conducted. The threshold level for adequate ecological representation may be expressed as, for instance, 10 percent of the maximum level of the historical range of variability for any ecological unit in the historical range of variability EDM. This percentage, when multiplied by the area within a habitat-type class, will provide a specific acreage threshold that needs to be provided to achieve adequate ecological representation. If an ecological unit in the matrix did not occur under historical disturbance regimes but occurs today because of past human influences, then the adequate ecological representation of that ecological unit would be 10 percent of zero, or none. In contrast, if the historical range of variability were found to include 70 to 90 percent of the acreage of a habitat-type class occurring in one ecological unit over an appropriate time interval, then 9 percent of the acreage in the habitat-type class would be the threshold level for that ecological unit. The 10 percent level is a hypothesized threshold level for adequate ecological representation that can be adjusted up or down based on further analysis and evaluation.

There are certain caveats to heed for this method to work effectively. First, it may not work for extremely rare ecological conditions within the planning landscape, such as bogs, that might consist of less than 1 percent of the overall acreage. Second, the threshold amounts may not be sufficient if provided in an artificial environment. In other words, the characteristics of other ecological communities surrounding the ecological units targeted for satisfying the adequate ecological representation threshold may need to contribute, at some level, to the overall habitat conditions for native species. The mix of ecological units providing adequate ecological representation must be further evaluated relative to the spatial distribution of each ecological unit across the landscape over time. This spatial analysis provides for the consideration of dynamic processes and allows for concerns such as linkages of metapopulations to be factored into the planning process. The species assessment discussed below provides a check on the adequate ecological representation of the coarse filter.

It should be stressed that the goal of adequate ecological representation is not a return to a historical range of variability. The goal is to identify threshold levels of each ecological unit required to maintain viable populations of all native species. This strategy allows for conservation of biological diversity as well as management flexibility within the planning landscape to meet other social and economic objectives. In this regard Haufler et al. (1996) stated: "Should these additional considerations result in a desire to reduce the area of a unit below the recommended threshold, all parties should have a clear understanding of the increased potential for future endangered species listings or losses from that planning landscape."

Check Adequate Ecological Representation with a Species Assessment

One assumption of adequate ecological representation as portrayed by the coarse filter is that minimum amounts of ecological communities properly distributed across the landscape will maintain viable populations of all native species (Haufler 1994). We recognize that the effectiveness of this coarse-filter approach may depend on relationships between species viability and amounts of ecological units that have not been well documented. For this reason, it is imperative that a check on the adequacy of the coarse filter be conducted (Haufler et al. 1996). This check can be accomplished through habitat-based assessments of species viability (Roloff and Haufler 1997).

Selection of the appropriate species is a primary consideration in conducting a meaningful check of the coarse filter. We recommend selecting species on the basis of the following criteria (Haufler et al. 1996):

1. Species dependent on ecological units that have undergone significant shifts in environmental conditions from those occurring under historical disturbance regimes.

2. Species with stenotrophic habitat affinities for specific ecological units.

3. Species with requirements for large areas that depend on specific ecological units of concern.

4. Species that represent the extremes in ecological unit differentiation (for instance, the assessment should emphasize species that depend on the vegetation conditions represented by extremes in vegetation growth stages and habitat-type classes portrayed in the EDM [Figure 7.2]).

The species assessment depends on a habitat modeling approach that provides a spatially explicit quantification of habitat quality and quantity (Roloff and Haufler 1997). One approach that provides these capabilities is "habitat suitability modeling" (Morrison et al. 1992). For each species, a habitat suitability index model can be identified or developed through the use of available information. The habitat suitability model provides a map of habitat quality and quantity based on vegetation structure, composition, and the organism's spatial requisites (Roloff and Haufler 1997). The results of model runs are habitat contour maps (Roloff and Haufler 1997).

The parameterization and mapping of vegetation conditions for each ecological unit provide the data necessary to run habitat suitability models. The adequacy of species-specific assessments depends on capabilities for mapping and describing habitats (Haufler et al. 1996). The EDM provides the classification for the habitat assessment. Land classification considerations for species assessments should include the quantification of vegetation structure and composition such that precise estimates of most habitat parameters (that is, within-stratum variation is less than between-strata variation) can be obtained (Haufler et al. 1996). The forested EDM for the Idaho Southern Batholith identified thirteen vegetation growth stages and eleven habitat-type classes. This level of resolution was perceived to adequately discriminate important habitat variables, represent the scale of habitat requirements for the majority of species of concern, and offer an operationally pragmatic delineation of ecological units. Thus, the foundation of realistic habitat assessments is a scientifically credible, ecologically based land stratification scheme (Haufler 1994). The mapping resolution of the EDM must also be at a fine-enough scale to minimize variance around habitat parameters (Haufler et al. 1996).

Addressing the viability of species across a landscape is a four-step process (Roloff and Haufler 1997). These steps include conducting and mapping the habitat assessment, determining evaluation criteria for the results of the habitat assessment, evaluating the habitat assessment relative to individual or pair viability, and aggregating the evaluation results into population viability planning objectives (Roloff and Haufler 1997). The aggregation potentially occurs at two scales, the subpopulation and population.

Minimum-quantity and -quality habitat objectives are identified for the species of interest using habitat suitability modeling (Roloff and Haufler 1997). Quantity and quality objectives link directly into population demographics—that is, high-quality habitat corresponds to high reproductive success, individual survival, and low mortality (Haufler et al. 1996). The habitat objectives should emphasize the maintenance of individual reproductive units (for example, a pair). The habitat contour map generated by the habitat model is evaluated by means of these habitat objectives (Roloff and Haufler 1997). The result of this evaluation is a map of home ranges, some of which are of sufficient quality to be viable (Roloff and Haufler 1997). Subsequently, individual home ranges are aggregated into subpopulations or populations, and the landscape is evaluated for connectivity among these aggregates (Roloff and Haufler 1997).

Once the population viability baselines are established, the amounts can be used to benchmark future ecological objectives (Haufler et al. 1996). If the coarse-filter approach indicates that these baseline conditions are provided over the planning horizon, it is assumed that habitat conditions for native species are being met (Haufler et al. 1996). The species assessments are used as checks against this assumption. Two conditions are evaluated with the species assessment. First, we ensure that the adequate ecological representation EDM is functioning properly. Second, the species assessment offers a mechanism to evaluate the spatial distribution of ecological units across the landscape.

As discussed in relation to determining a desired future condition EDM, successional or tree growth and yield models can be used to project ecological units into the future. This activity may be complicated by the presence of new influences on either overstory or understory successional trajectories, including the presence of exotics, livestock grazing, or altered levels of herbivory by native species. Resultant shifts in ecological-unit acreages can be monitored by means of the EDM and geo-referenced with the GIS (Haufler et al. 1996). Species assessments can be conducted over time, and changes in habitat conditions can be monitored to evaluate ecological objectives. Thus, coarse-filter assessments with species checks can occur throughout the planning time frame. The process incorporates the structural and spatial dynamics of habitat, recognizing that habitat patches can move in response to changing environmental conditions (Haufler et al. 1996). The species assessment provides a means of evaluating the effects of these dynamics on species viability.

Identify Social and Economic Concerns

This chapter has heretofore focused on the ecological objectives of ecosystem management, specifically the conservation of biological diversity. We recognize that successful ecosystem management also incorporates social and economic objectives and, in doing so, increases the likelihood of achieving the ecological objectives. Social and economic conditions are extremely dynamic and operate at a multitude of spatial scales. Small-scale economic and social issues (such as those of towns) typically exhibit strong relationships to the ecological description of the planning landscape. For example, the economics of communities dependent on logging exhibit a direct link to the ecology of an area. Similarly, the social values of individuals associated with these communities typically reflect their dependency on natural resources. In contrast, large-scale economics and social values (such as those of states) are typically driven by a suburban populace with completely different views on their relationship to the ecology of a landscape. Thus, successful integration of economic and social values into an ecosystem management process requires a structured, hierarchical approach, much like the hierarchical method used in the ecological assessment.

There are multitudes of economic and social considerations in ecosystem management. The methods for quantification of all these considerations are beyond the scope of this chapter; however, some of the ecologically based tools can be used to characterize economics and social values. For example, estimates of timber yields can be linked to ecological units, and, thus, the EDM offers a means of parameterizing timber values and projecting yields. Nevertheless, it is important to note that many economic factors are determined at scales completely unrelated to the planning landscape (for instance, the ecological assessment of a planning landscape does not influence regional market conditions or nationwide supply and demand)—thus the importance of considering economics at a multitude of scales.

Some social values can also be quantified by means of ecologically based tools. For instance, ecological units in the EDM that have high aesthetic or cultural values can be identified as a social concern. Much like economics, social concerns often develop and occur at scales independent of the planning landscape and the ecological assessment. Thus, broad-scale social values can contradict the ecological objectives established for a planning landscape. Nevertheless, social and economic concerns must be addressed to develop effective ecosystem management plans. The challenge is to provide a mechanism for integrating all three types of objectives. This process provides that mechanism by means of the EDM and adequate ecological representation for the planning landscape.

Although the majority of this chapter emphasizes the ecological considerations of ecosystem management, the importance of social and economic issues should not be considered minor. A thorough understanding of all three components is essential to properly define desired future conditions.

Develop Desired Future Condition EDM

The challenge of planning for ecosystem management is to meet complex and diverse objectives from the same landscape. Ecological objectives have been among the most elusive to address in a systematic and effective manner. The process presented here allows for ecological objectives to be quantified for a planning landscape. Under this approach, the identification of acreage of each ecological unit needed to provide minimum threshold levels of adequate ecological representation becomes a base level that should be maintained for an effective coarse filter to be provided. The EDM can be projected to future times by modeling successional change within each habitat-type class and projecting future conditions by understanding transitional probabilities among the ecological units that make up a habitat-type class. One desired future condition, if ecological objectives are to be met, is to provide for adequate ecological representation at all times in the planning horizon. Thus, the adequate ecological representation EDM becomes one set of minimum conditions to be provided.

Social and economic objectives also need to be addressed in the planning process. Identifying desired future social and economic outputs and developing consensus on their correct balance are the challenge of cooperative efforts. While many social and economic objectives cannot be directly represented as an output of an EDM, the ability of the landscape to have the correct mix of conditions to meet various social and economic objectives can be evaluated. Some social and economic objectives may be spatially limited, constrained to specific locations in the landscape. Where this occurs, the ecological units making up these specific locations can be designated for these uses and their acreage so designated within the EDM. For selected natural resource commodities, from timber to mushrooms, ecological units can be evaluated and ranked as to their potential supply. While certainly different views on desired conditions will occur, this process allows for ecological objectives to be quantified and a framework for addressing social and economic demands identified.

Because of the complexity of views on desired future conditions for social and economic objectives, it may not be possible to determine an exact desired future condition EDM. However, it is highly likely that some desired conditions, beyond those meeting adequate ecological representation, can be identified and quantified. To the extent possible, a desired future condition EDM can be developed and accommodated in the planning process. While this EDM may go through frequent revisions as social or economic demands change, it can still provide a critical planning tool for meeting all three types of ecosystem management objectives.

Develop Existing Condition EDM

An understanding of existing ecological conditions is necessary to determine baseline conditions from which management actions will be developed. The EDM provides the framework for describing and quantifying existing ecological conditions. The purpose of this step in the process is to quantify the entire planning landscape as to its current ecological communities. An existing condition EDM is produced by quantifying the total area of each ecological unit occurring in the landscape.

The landscape can be mapped for ecological units of the EDM by using habitat-type models in tandem with remote sensing of vegetation growth stages. However, the resolution of this type of map is fairly coarse and the accuracy questionable. Conducting landscape planning from such a map will produce uncertain results until resolution and accuracy problems have been evaluated and corrected. Therefore, a better approach is a collaborative effort designed to produce compatible land classifications and analysis (Haufler 1995) coupled with cooperative planning and management. This approach offers each landowner the capability to evaluate his or her contributions toward the ecological objectives of ecosystem management in comparable, quantifiable terms (for example, hectares of ecological units), in addition to providing knowledge of existing habitat conditions to evaluate the contributions landowners can make toward species viability. The primary challenges to ecosystem managers attempting to quantify existing ecological conditions will be the development of effective cooperative partnerships that will provide incentives for landowners to participate and that will reduce the fear of regulatory repercussions resulting from increased knowledge of ecological conditions.

Identify and Implement Required Actions

To identify required actions on a landscape, the existing condition EDM must be quantitatively compared to the desired future condition EDM. Where existing ecological units are underrepresented in the desired future condition EDM, management plans should emphasize actions that potentially provide the necessary amount identified within an acceptable time frame. Implementation actions will consist of changing one vegetation growth stage to another within a column or habitat-type class of the EDM. Haufler et al. (1996) discussed several actions that could produce a desired future condition. However, each individual landscape will dictate the appropriate mechanisms for achieving the desired changes in vegetation growth stages.

In the process of working toward achieving desired future conditions, an additional concern is the spatial and temporal arrangement of ecological units within the landscape. This may be important to providing quality habitat at the home-range scale or connectivity among populations at the landscape scale. In either case, species assessments are helpful in identifying a spatial arrangement of ecological units that minimizes viability issues for species of concern. All spatial arrangements of ecological units must also incorporate a temporal context. It is imperative that all required actions respect the dynamic nature of landscapes (Camp et al. 1997) and provide for changes in the spatial distribution of habitat. The concept of permanent habitat configurations is not compatible with ecosystem management objectives on dynamic landscapes and will not conserve biodiversity in the long term.

Implementation must also consider the objectives of individual landowners when one is evaluating their contributions to desired future conditions. The identification and implementation of actions must consider social and economic concerns of both the general public and landowners. The challenge to ecosystem managers will be achieving the right spatial distribution of ecological units that will maintain the landscape at or above the threshold of adequate ecological representation while also providing opportunities for optimum social and economic benefits (Haufler et al. 1996).

Monitor, Evaluate, and Adjust

All efforts at integrating biodiversity objectives into ecosystem management at a landscape scale are relatively new and represent conceptual thinking about how to accomplish this goal (Haufler et al. 1996). A strength of our process is that it provides a mechanism for quantitatively measuring progress toward a desired future condition that considers biodiversity. The approach described in this chapter represents a data-intensive, ambitious undertaking (Haufler et al. 1996); however, the inherent complexity of ecosystems dictates this type of effort. The next steps are to test the various components and to adjust the process accordingly.

Monitoring and evaluation of this ecosystem management process are also critical for continued improvement. We recognize that some data and knowledge gaps are imbedded throughout the process. But even though data gaps exist, it is imperative that resource professionals move ahead with implementation of conservation efforts for biological diversity. The process we have presented in this chapter is based on the best available science, is intuitively logical, and is communicable to on-the-ground land managers. The process will undoubtedly be improved and refined through the adaptive management process (see Nudds, Chapter 11), which certainly justifies the necessity for a structured monitoring and evaluation program.

Chapter 8

Maintaining Biodiversity in an Intensively Managed Forest: A Habitat-Based Planning Process Linked with a Fine-Filter Adaptive Management Process

William A. Wall

Public demands drive markets for forest commodities. But in addition to its demands for forest products, society also wants to maintain biodiversity and functioning riparian and aquatic systems, desires that it expresses through federal and state regulations such as the Endangered Species and Clean Water Acts. These needs, coupled with recent advances in geographic information system software technology and an emerging emphasis on managing at the landscape scale, have led to the development of multiple ecosystem and landscape management approaches.

As various land management agencies and private forest industries initiate landscape-scale planning efforts, it is important to recognize the validity of initiating and testing multiple approaches and opportunities to work with differing ownership and management objectives to develop optimal solutions.

For the landowner engaged in timber production, the primary objective is efficient, cost-effective yields and the ability to continue to conduct business. Through systematic planning processes that integrate adaptive management, much can be achieved in intensively managed forests to maintain biodiversity and ecosystem functions, thereby sustaining production and encouraging favorable public opinion.

This chapter describes a practical adaptive approach to habitat-based landscape planning that works to achieve both goals. The process is linked with a fine-filter species assessment and monitoring process. This chapter does not define the ideal, large-scale ecosystem management process but describes a practical approach to planning, implementing, and monitoring a landscape strategy for the maintenance of biodiversity in private industrial forests.

Many ecosystem management approaches are top-down: information collected at a macro scale with coarse resolution is then applied to smaller scales. The approach described here is bottom-up: information is collected at a fine resolution and aggregated to larger scales. The described approach could be integrated with additional efforts of other landowners at several scales or with a large, coarse-filter approach to achieve a macro-scale or ecosystem-based collaborative process. In addition to its flexibility, another benefit of this strategy is that it provides a continuous planning and monitoring process integrated with information management, modeling, and monitoring.

The Planning Process—Components and Goals

An integrated planning process for private industrial forest lands will have numerous goals. At an ecological level, it should account for wildlife habitats and functioning aquatic systems across landscapes. It should also integrate short- and long-term business goals for annual production and economic goals. To meet these needs, the key ecological components required are information on ecosystem capability, potential vegetation, disturbance regimes, and the functions of aquatic and riparian systems. This ecological information must be integrated within the same system with timber inventories, growth and yield projections, production optimization, and harvest scheduling. The ecological and production information must then be linked with an understanding of expected species distributions, habitat requirements, and the current condition of vegetation and aquatic systems.

From an industrial management perspective, the primary goal is low-cost maintenance of a sustainable level of forest products and production while meeting regulatory requirements (Boling et al. 1996). Many would also argue that the industrial landowner has an obligation to meet certain societal expectations for stewardship of land, including maintenance of biodiversity. Some industrial landowners have responded to this challenge by developing landscape-planning processes. Creating a planning process that addresses long-term fiber production, short-term economic goals, and functional habitat for native species requires integrating and managing significant amounts of information. Critical to this undertaking is the ability to analyze the current condition of the landscape and to predict the future landscape created through management actions.

Predicting a desirable future for a landscape is an optimization problem that originates with the landscape's current condition and considers the maintenance of an acceptable level of sustainable timber production and a diverse array of quality terrestrial habitats and riparian and stream function. A modeling process that predicts future landscape conditions in terms of fiber production, habitat condition (structurally and spatially), and riparian and stream impacts is required in order to optimize plans for meeting all of these objectives.

The critical component of a complete process that is rarely or fully integrated in continued planning is the adaptive management feedback loop, which uses past experience to improve future management decision making. Another key component is how information drives this dynamic process and allows management flexibility, continuous updating, and refinement of annual or multiyear projections.

So, therefore, we can summarize the goals of this process as follows: A key reason for an industrial landowner to invest in the development and monitoring of a landscape-planning approach is to maintain high standards for land stewardship and to gain public acceptance of intensive forest management practices while remaining economically competitive. A primary ecological goal of this approach is to select and schedule activities that maintain critical landscape components and sustain ecological processes. Additional goals are directed toward maintaining specific landscape habitat components and functions or guidelines for specific management activities. The general wildlife goal is to manage a diverse array of habitat components within stands and across landscapes to maintain native vertebrate species richness, special management needs for sensitive species, and habitat requirements for featured game species. The purpose of this goal is to contribute to overall viability of native species within a greater landscape, not to maintain viable populations within each of the planning units. The aquatic habitat goal is to establish fully functional riparian management areas as a distinct and stable component of the landscape with minimal disturbance and to achieve full stream function. All of these goals are integrated within the analysis, planning, and monitoring components of this landscape management process.

General Description

Under this approach, information is collected and aggregated at four spatial scales. The primary scale for inventory is forested stands ranging in size from 25 to 400 hectares and averaging about 200 hectares. Potlatch maintains an inventory of the current structure of each stand and the size, age, density, and species of trees within it as well as parent material, habitat type, digital elevation model (DEM) attributes, and climatic attributes, such as annual precipitation and solar insulation.

Stream and riparian management objectives in subbasins of 5,000 to 12,500 hectares require the ability to aggregate stands to this second spatial scale, where watershed analysis is conducted. These subbasins are aggregated to the third scale of 50 million to 125 million hectares or Landscape Management Unit (LMU). Landscape plans are developed, implemented, and monitored at the LMU scale. Plans and LMUs can then be aggregated to larger areas within the ownership as needed or shared at a fourth regional scale if needed for specific wide-ranging species, such as elk or lynx.

The only top-down aspect to this planning process is a linear optimization planning program that allocates forest stands for harvest scheduling across the entire ownership (all LMUs) simultaneously. Thus, timber harvest is optimized across the entire potlatch ownership of 245,835 hectares not within the LMU. The initial allocation of a harvesting schedule for each of the LMUs is driven primarily from a forest production perspective. These allocations are on a temporal scale of eighty years, approximately one average rotation. Maps showing predicted habitat are produced at ten-year intervals. Initial allocations can then be adjusted based on habitat objectives for any of the LMUs.

The planning process is a continuous dynamic process linked with continuous monitoring. The stand-harvesting allocation program is run on an annual basis. However, a six-year cycle is required to develop plans for over fifty LMUs on 1.68 million hectares. Approximately nine are completed each year. Once an initial plan has been developed, it is revised on a continuous six-year cycle with new information on resource conditions and the adequacy of management objectives developed through the monitoring process. The initial plan serves as a baseline, but it has the flexibility to deal with management needs as they arise, because continuous monitoring allows the development of trend information on resource conditions. In addition, habitat association models are used to predict species changes over each LMU. If needed, these models can be run on several LMUs that have been aggregated.

Planning Steps

The planning process is composed of five progressive steps that, when diagrammed, form a continuous planning and information feedback loop. These steps are intended to be continuous. The planning and monitoring process is never completed but is cyclical. The steps are as follows:

1. Current condition analysis. This step describes the current status of the LMU in terms of timber production, riparian and aquatic condition, structural stages of the stand, distribution and wildlife species distributions (both through monitoring and empirically based models), and climatic and ecological background information such as site index, habitat typing, and physical site classification.

2. Establishment of objectives and constraints. Specific objectives and constraints are established for each LMU to integrate business objectives with goals for wildlife habitat and aquatic and riparian management.

3. Plan projection. The plan is projected through growth and yield forest modeling (a derivation of Prognosis (Wykoff et al. 1982) to produce a time series of spatially specific ten-year projections of expected forest structures for eighty years. The predictions of future forest conditions can then be used to forecast future wildlife habitat quality and species distributions and, in some cases, through the application of locally developed habitat association and density models, potential density estimates for various species using locally developed habitat association and density models.

4. Plan implementation. The plans are implemented through one-year and five-year budgeting. Plans are flexible and may be changed during the budgeting process on an "as-needed"

basis. Although not currently part of the process, the plans will potentially allow field personnel to analyze and make changes quickly. Any plan changes are incorporated in the next LMU plan review and update at six years.

5. Plan monitoring. The process involves three types of monitoring: implementation, effectiveness, and model development and validation, each integrated with the LMU plans.

Description of Planning Components

Current Condition Analysis

Current condition analysis focuses on several areas, including timber production and silvicultural treatment of stands, watershed analysis, vegetation composition, structure assessment, wildlife habitat, and species distribution. This chapter will not address the timber management component.

Watershed Analysis

The purpose of watershed analysis is to set baseline conditions, determine potential management hazards and basic watershed potential, develop site-specific BMPs, and understand opportunities for restoration of any of the watershed's functions. From this baseline and analysis, site-specific BMPs are developed for streams, riparian management zones, and road systems. Analysis also promotes knowledge of underlying hazards for management in the watershed. Hazards include risk ratings for potential landslides and road surface erosion.

Road, riparian function, and stream condition assessments establish the baseline of and indicate causes for current conditions; in turn, they suggest appropriate management measures.

- Road assessments include risk analysis for landslide hazards and surface erosion rates, the current condition of road systems, and specific maintenance needs.

- Riparian function analysis includes assessment of shade, bank stability, sediment filtration, and recruitable large organic debris.

- Stream condition analysis includes temperature, sedimentation, structural diversity of the habitat (pools and riffles), stream bank integrity, and fish and invertebrate distributions and densities.

Forest Structure Classification

Many attributes within the structure of forest stands relate to habitat quality for various species. Forest structure classification (see Table 8.1) is based on the following attributes: (1) number of layers (including understory); (2) snag, cull size, and densities; (3) dead and down woody debris; and (4) size and density of overstory. Forest structures have been classified in the past through the use of succes-

Table 8.1. Within-stand Structure Components, Size Ranges, and Target Quantity for Stand Management of Structural Habitat Diversity

Structure Type	Size Range	Future[1] Rare Components	Target or Ideal Quantity
Snags	10+ in. high 10+ ft. DBH	20+ ft. DBH 40–75 in. high Heart rot	4–6 per acre Scattered or clumped
Culls	16+ ft. DBH	18+ ft. DBH	2–4 per acre
Wolf trees	Broken top or full tree	Broken tops, forked tops, heart rot	Scattered or clumped
Green residuals	4–8 ft. DBH		Dominant and whips
	9+ ft. DBH		Scattered or clumped
Dead and down logs	12+ ft. DBH 10+ in. long	20+ ft. DBH 40+ in. long (Rotten/ unburned)	4–8 per acre Scattered or piled
Layers[2]	Shrubs (brush understory); 0–3 ft. DBH understory (regeneration); 4–8 ft. DBH midstory; 9+ ft. DBH overstory		Shrub layer in mature stands[3] 50+% ground cover
Patches	1/4–1/2 acre Layered or not Consists of any layer		1–3% of stand

1. Rare habitat component within intensively managed stands.
2. Amount of vertical structure within stands must be tempered by silvicultural goals.
3. Shrub and regeneration layers in stands with overstory (9+ ft. DBH) are a primary target.

sional stages. Since typical successional patterns are constantly disrupted through forest management practices, it is now necessary to use a forest structure classification.

Many approaches for categorizing structure within managed forest stands (O'Hara et al. 1996) have been developed. These structures allow the prediction of species associations as well as water yield effects within watersheds. Structure forms the basis for habitat-quality ratings, mapping of habitat both spatially and temporally, and use of habitat association models.

Wildlife Habitat

Habitat management in the context of intensive forest management must be planned and conducted. There is no practical way to plan for the complexity of habitat conditions needed by the approximately 400 vertebrate species within the inland Rocky Mountain biome. Among the many approaches to this problem are managing for indicator and umbrella species and using a coarse-filter community management approach (Haufler et al. 1996).

The approach used is a life-form matrix that assigns species to a specific lifeform (Thomas 1979). "Life-forms" are groups of species that exploit similar major habitat components for reproduction and feeding. The life-form matrix allows a specific focus on key habitat components at a stand-level scale. However, in order for managers to integrate biodiversity goals into planning and actual management implementation, the life-form matrix needed to be simplified and specific target habitat components developed at several scales. Through this form of the life-form matrix approach, species associated with major habitat components are classified into broad categories and, when based on influences of forest management, can be aggregated into management guilds. The primary assumption is that if these major habitat components are managed appropriately, management guilds will be positively affected.

This assumption, of course, is as broad as the concept of management guilds. Each species within a management guild will respond differently to disturbance regimes (forest management) and the resulting stand structures, patch sizes, and juxtaposition. Generalized management planning produces high- to low-quality habitats for species within the same management guild. Across a large area and through time this approach will maintain a diverse array of key habitat components within an intensively managed forest land-scape.

The key to validating the effectiveness of this approach in main-taining species presence is an integrated approach to monitoring and adaptive management. Table 8.2 lists initial management guilds with target habitat features and the number of species associated with each life-form. Management guilds are considered through the continuous planning process. The initial current condition analysis describes the status of the various habitat components; then a plan to maintain or enhance these features is developed. Monitoring of these features is accomplished at two spatial scales: at the stand scale through implementation monitoring, and at the LMU scale through updated analyses of future and current conditions.

Table 8.2. Potlatch Habitat Management Guilds as Developed from Life-forms Described by Thomas, 1979

Potlatch Management Guilds	Life-form Designation	Number of Species	Primary Structure
1. Aquatic	1	Fish, amphibians	Standing or running H_2O
2. Riparian	2–3	54 herps, small mammals, birds	Habitat adjacent to streams, ponds, wetlands
3. Rock habitats	4	32 birds, bats, herps, and small mammals	Cliffs, caves, rim-rock, talus
4. Ground associates	5–6, 15–16	55 small mammals, 50 birds	Dead and down logs; brush, forbs, burrows, and banks
5. Brush associates	7–8	36 birds	Stand understory brush and perch sites
6. Mature trees and interior forest stands	9–12	49 birds, small mammals	Large deciduous and conifer trees in 100± acre patches
7. Cavity users	13–14	50 birds, small mammals	Snags and heart rot culls

Objective Setting

Setting specific management objectives always requires a trade-off. Under our approach, objectives are set for each landscape by simply determining how to meet all of the key overall goals and then developing a plan that maintains or enhances those goals over the planning period. Thus, depending on current conditions, some areas will need active restoration, and others will need sensitive management or protection.

A good example is riparian zone management. In some cases, historical management has created a current condition that requires active restoration in order to reach full function. In others, sensitive management is required to maintain or enhance full function. In the case where previous management has left few larger snags for primary cavity nesters, silvicultural practices would be applied to enhance retention and development of appropriate numbers and sizes of snags. In addition to an activity schedule for performing forest management tasks, key objectives with a time schedule for completion are developed.

Plan Projection

The LMU plan is projected through "growing" forest stands with a growth and yield forest model and converting stand tables into structure classes as described above. These structures are then mapped and graphed for spatial juxtaposition analysis and analysis of quantity of various stand structures. The forest growth and yield models are augmented with understory vegetation models for predicting understory structure and composition, following forest management disturbance. Empirically developed habitat association models (currently for sensitive amphibian and bird species) are used in conjunction with the forest growth and structure models to determine the range of quality habitats being maintained throughout the plan period for a specific LMU. For instance, if the amount of interior mature structurally diverse forest is reduced to below one-third of the LMU area for more than one ten-year period, then changes in timing and spatial distribution of harvest units will be made and the projection modeling is redone.

Plan Implementation

Plans are implemented through a one-year and five-year budgeting process. These budgets are updated on an annual basis. Changes in the LMU plan relative to location and timing of harvest can be made "on the fly" based on business objectives and need for flexibility because of weather and other implementation constraints. However, these changes must be modeled, and changes must meet the overall objectives of the plan. The next review, at six years, allows for a complete update and reevaluation of the LMU plan.

Implementation Monitoring

This type of monitoring is similar to that used in a continuous improvement or quality-control process for many business or manufacturing processes, ensuring compliance with predetermined standards. If standards are not met, corrective measures can be taken. This process component is typically initiated through a training program for employees and contractors. The training program explains the standards, performance measures, and implementation monitoring process. Continuous improvement occurs as feedback from monitoring all management activities is used to hold employees and contractors accountable for performance. These standards and measures can apply to any type of management, including silvicultural standards, to meet growth and yield requirements and goals for structural diversity of habitats.

Although implementation monitoring feedback is an important component of improving performance and plays a role in adaptive management, this process does not generate the information necessary to improve (adapt) the effectiveness of implemented standards. Standards for wildlife habitat are established through the components targeted by the management guild approach. Implementation monitoring allows for feedback about whether the appropriate targets are being met at the stand scale. Landscape components are checked through the continuous planning process by updating the current condition analysis for each LMU.

Adaptive Management

A key to the described landscape planning process is that it is truly an adaptive management program. Critical management elements, such as riparian function, have continuous monitoring and evaluation, supporting the ability to continuously update landscape plans with new information. Monitoring is conducted in such a way as to continuously update species distributions, test and enhance habitat association models, and determine the effectiveness of management guidelines at maintaining and enhancing specific criteria for functioning habitat.

Riparian and Stream Monitoring

The effectiveness of a monitoring program relies on key management guidelines that have been implemented to maintain full riparian function. The question is whether the management activities, as dictated by guidelines, result in accomplishing specific, predetermined LMU objectives. For example, LMU objectives are accomplished for riparian and stream function by establishing baseline conditions within an LMU and then monitoring on a six-year basis. Annual variation is tracked in several LMUs that contain sensitive or rare species. This approach to effectiveness monitoring, termed "atlasing," allows 80 percent of the monitoring effort to be applied to establishing initial baselines and subsequent long-term trends and 20 percent of the effort to determining short-term trends and effectiveness by monitoring annual variation of both habitat condition and certain key (sensitive) species. Established management standards that are monitored include riparian and stream management, road building and maintenance, and silvicultural practices producing diversity of stand structure.

Bird Monitoring

The goal of the bird-monitoring program is to establish habitat relations among landbirds at several scales, from stands to LMUs, and to serve as a check for refining and identifying management guilds with increasing precision. Predictive models of habitat use are developed for specific forest structure classes. These models are used to predict bird assemblages both spatially and temporally in LMUs in the landscape-planning process. If significant reduction in habitat for any species is noted during the plan projection process, changes in the plan can be made before implementation begins. These first-generation models are validated by predicting expected assemblages within an LMU, then monitoring to test the predictions. This is the fine-filter component of this habitat-based planning process.

Amphibian Monitoring

The amphibian monitoring program was initiated to develop habitat association models for sensitive amphibian species associated with riparian and aquatic habitats and to serve as a check on the management guild approach. The monitoring process is an atlasing approach. Initial monitoring is conducted at the LMU watershed scale. The objectives are to develop predictive habitat association models and distribution maps and to monitor annual variations in key populations. These objectives are accomplished by establishing several breeding sites to be revisited on an annual basis (about 20 percent of the annual amphibian monitoring effort). The other 80 percent is used to search new LMUs to establish the presence and absence of habitat association models and to continue to test and refine them. By revisiting each LMU on a six-year basis, the long-term persistence of populations within LMUs can be monitored by a manager.

Summary and Conclusions

The described landscape management planning and implementation process is a truly integrated adaptive management system. The collection of data at the stand level allows a high level of accuracy and precision and the ability to aggregate to larger scales. The system is driven by information that is constantly updated. Plans are dynamic and flexible and are reevaluated on a six-year schedule with new information. The process is fully integrated with business and commodity production planning. The monitoring feedback loop accomplishes two objectives: (1) implementation monitoring checks whether management guidelines and plans are being conducted properly; and (2) effectiveness monitoring checks whether the guidelines are effective at accomplishing the expected habitat results while simultaneously enhancing habitat association models, species distribution information, and an understanding of annual variation in select processes and species populations. This process continuously informs itself and managers.

There are two primary weaknesses in this management planning system. Without a linkage to regional-scale analysis, this process lacks regional context to set priorities for species and key habitat components. However, the process does have sufficient flexibility to easily incorporate information from larger-scale analysis efforts. Although this system incorporates key ecological processes such as watershed integrity, it is not adequate at determining such ecological perspectives as "adequate ecological representation" (Haufler et al. 1996) or historic natural range of natural variation (Morgan et al. 1994). Both of these methods of setting landscape context and goals could be applied within the framework but are not currently considered. In addition, this process does not use expected or potential successional pathways. Instead, the framework uses growth and yield forestry models with the expected intermediate silvicultural treatments like the primary expected disturbance. Thus, the process does not link to expected "natural" pathways but does more accurately reflect actual expected disturbance regimes and offers the ability to constantly update the system.

C h a p t e r   9

Landscape Planning and Management in Agro-Ecosystems: The Canadian Prairie Experience

Terry G. Neraasen and
Jeffrey W. Nelson

This chapter describes the habitat conservation activities of Ducks Unlimited Canada as an example of ecosystem and landscape management. Ducks Unlimited is an international, not-for-profit organization dedicated to the increase and perpetuation of North American waterfowl populations through the conservation and management of key habitats throughout the birds' annual cycle. Since 1938 more than 13,000 pieces of land and nearly 1.2 million hectares of habitat have been protected and restored in Canada for the benefit of waterfowl and other wildlife species.

The focus of conservation programs has always been in the Prairie Pothole Region (PPR). Eighty percent of the work has occurred in the Prairie Provinces. Millions of small, shallow wetlands dot this approximately 750,000-square-kilometer region, formed 10 to 15 thousand years ago as the Wisconsin period of glaciation ended (Stewart and Kantrud 1971), making it of well-known importance to breeding waterfowl. Batt et al. (1989) estimated that, on average, half or more of the surveyed populations of many waterfowl species breed in the PPR, including mallard (*Anas platyrhynchos*), northern pintail (*A. acuta*), blue-winged teal (*A. discors*), American wigeon (*A. americana*), gadwall (*A. strepera*), northern shoveler (*A. clypeata*), canvasback (*Aythya valisineria*) and redhead (*A. americana*), and ruddy duck (*Oxyura jamaicensis*). Of these, mallard, northern pintail, blue-winged teal, and American wigeon are species of special concern because populations throughout the 1980s and early 1990s have been considerably below population goals established by the North American Waterfowl Management Plan (NAWMP) (Anonymous 1986). The NAWMP, officially signed in 1986 between the governments of Canada and the United States, with Mexico becoming a full partner in 1994, has established cooperative international efforts to reverse the declines in waterfowl populations and their habitats. In addition, the lesser scaup (*Aythya affinis*), which also makes extensive use of the PPR for breeding and staging, is a species of concern. Although recent survey information indicates that only northern pintail, American wigeon, and lesser scaup remain below NAWMP goals (Anonymous 1994a), given that populations fluctuate considerably in response to a variety of factors, including habitat conditions, all remain of concern to wildlife managers.

Populations of these PPR-nesting species are thought to be most limited by poor breeding success, which is influenced by breeding effort, nest survival to hatch, duckling survival to fledging, and hen survival during the breeding season. Water and wetland conditions strongly influence the degree to which waterfowl settle into the PPR to breed, breeding effort as measured by numbers of hens that attempt to breed, and renesting effort of those that do. These factors, along with the fact that large areas of intact grassland were still available, allowed high production of waterfowl in the PPR during the 1960s and 1970s. Good wetland conditions during the spring arrival period usually resulted in good brood production and survival because the two most important factors, nest survival and duckling survival (brood success) (Johnson et al. 1992), were generally assured. In recent years, rates of nesting success for upland nesting ducks in the PPR have mostly averaged less than 15 percent, the minimum thought to be required for population stability (Klett et al. 1988). In addition, summer mortality rates for nesting mallard hens in excess of 30 percent and 50 percent duckling mortality have recently been documented in intensive studies of mallard populations in the PPR (Sargeant and Raveling 1992).

Poor waterfowl and other wildlife productivity and survival in the PPR are related to significant ecosystem alteration and fragmentation. Approximately 50 percent of the wetlands present in the PPR prior to settlement have been lost to agricultural drainage and conversion to other uses. In some areas, 90 percent have been lost. Of those that remain, 70 to 90 percent are modified by cultivation, burning, or brushing in· any given year. Agricultural conversion of native prairie grasslands and wetlands has been most intense in the parkland and tallgrass prairie, ecoregions where soils are rich and rate of evaporation at most equals precipitation (Trottier 1992). In areas farther west, where arid conditions and poorer soils exist, cattle ranching is extensive, so large tracts of prairie remain. However, such grassland areas are often extensively grazed, resulting in poor nesting habitat and severe disturbance during critical nesting periods for both waterfowl and other ground-nesting species. Greenwood et al. (1995) indicated that nest success decreased by about 4 percentage points with every 10 percent increase in cultivated land in their study of duck nesting in prairie and parkland areas of the Canadian PPR.

Human disturbance and manipulation have also altered predator communities. In eastern portions of the PPR, where cropland is extensive, predator communities are diverse and dominated by the red fox (*Vulpes vulpes*), which is known to be particularly damaging to upland nesting waterfowl and likely other ground-nesting birds (Sovada et al. 1995). In areas where grasslands are more extensive and coyote (*Canis latrans*) populations are dominant over those of foxes, waterfowl production is generally better (Sargeant et al. 1993). Compared to eastern areas, loss of nesting hens and eggs tends to be much lower in western grasslands because coyotes exclude red foxes and fewer predator species are present.

Populations of a variety of nonwaterfowl species, especially ground-nesting passerines, have declined over the past several decades in the PPR, likely as a result of habitat loss and degradation (Anonymous 1994b, 1996) and resultant effects on nesting success and brood survival. Problems faced by these species are similar to those faced by waterfowl, and solutions proposed for either will likely benefit the other.

Any attempt to address these problems must focus on securing and enhancing wildlife habitat on large areas of the PPR. Fragmentation of ecosystem components must be reversed so that systems become integrated, functional units.

Program Planning and Delivery

The NAWMP (Anonymous 1986) clearly stated that if its goal of returning declining waterfowl populations to the average numbers that occurred during the 1970s were to be achieved, large areas of habitat would have to be protected and improved. The PPR was highlighted as a "Key Waterfowl Habitat Area of Concern." In Canada, the Prairie Habitat Joint Venture (PHJV), sponsored by government and private agencies in provinces of Manitoba, Saskatchewan, and Alberta, was put in place to provide the administrative structure to plan, fund, and deliver the habitat and other programs necessary to address problems faced by waterfowl in the region.

Conservation program planning and delivery under NAWMP are based on large units or landscapes in which a variety of habitat protection and restoration activities are delivered as an integrated whole. In order to increase waterfowl recruitment, there is a need to maximize the quantity and quality of grassland or similar upland cover and to maintain and restore wetlands. This must be carried out in areas with high density of wetlands so that when water conditions once again improve, nesting cover for waterfowl and other ground-nesting species will be available. Programs aimed exclusively at wildlife production, as well as more extensive programs that promote wildlife-friendly agricultural practices, are necessary if enough of the PPR land base is to be improved to allow the improvements in waterfowl production. Some limited research has shown that agricultural practices such as use of no-till winter wheat have the potential to be significantly more productive of ducks than conventional tillage practices (Duebbert and Kantrud 1987). If large areas of cropland could be converted to direct-seeded fall crops, the small improvements in nest density and success spread over large areas of historically high duck production could have significant impacts on overall waterfowl populations.

Principles

From a biological perspective, planning and implementation of conservation programs are driven by a number of principles aimed at ensuring that the fundamental ecosystem perturbations affecting wildlife productivity are addressed. To be effective, attempts to restore waterfowl and other wildlife productivity must

- protect remaining grassland areas and increase their extent and distribution so they form large blocks of habitat with maximum wildlife potential

- provide for linkages or connectivity between various important ecosystem elements so that ecological integrity and functionality are restored

- increase diversity of habitat types and components of individual types

- increase breeding populations through increased survival and recruitment

- focus on regions of quality, high-density pothole complexes within the Prairie Provinces ("priority areas")

- use the best of the best protection and restoration efforts within each priority area

- be based on habitat goals that maximize waterfowl population goals

- provide planning and delivery to landscape units that are biologically distinct

- incorporate multiple-species values into planning, program delivery, and management

- consider occupation and use by people as integral to the system

The last principle is particularly important in that the nature and magnitude of the program must be realistic and achievable. The habitat plans need to balance effectiveness in achieving wildlife and ecosystem objectives against what will ultimately be acceptable to owners of the land, because the majority of the most important and altered land base is privately owned. Restoration of ecosystem function must allow coexistence of wildlife and human activity. What follows is a description and discussion of how this has been approached over the past decade in Prairie Canada.

Initial Targeting and Landscape Delineation

Personnel of the PHJV initially selected priority areas for program delivery by ranking townships (approximately 92-square-kilometer units) based on the numbers and permanence of wetlands. Within the PPR in Canada, approximately 560,000 square kilometer units have been classified for wetlands within approximately 684 1:50,000 National Topographic System map sheets covering most of the southern portion of each of the provinces (Figure 9.1). Wetland statistics from this classification provided a valuable first step in the targeting of habitat programs. In addition, approximately 291,000 square kilometers of the same area have been classified as land cover and land use, providing a database for delineating baseline conditions within areas that are targeted for protection and restoration.

The initial step in targeting involved the segregation of groups of townships that ranked high in both density and permanence of wetlands. These groups formed the original large "key program areas" (KPAs) in the prairie and parkland biomes in all three provinces. Eleven such large target areas were initially delineated in the prairie grasslands and parklands, all of which are characterized by the presence of morainal pothole complexes known to be important waterfowl breeding and staging areas.

Further targeting within each of these KPAs delineated delivery areas or target landscapes. Wetland rankings at a finer scale (section and quarter-section level) were used as the primary factor at the landscape level, helping to pinpoint areas that would become the focus within KPAs for program delivery. Sections that ranked highly for

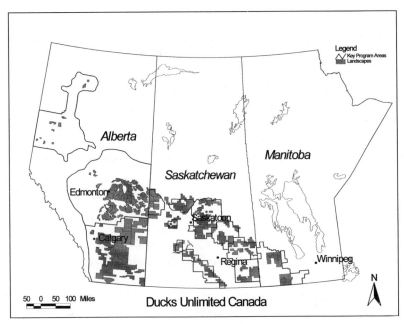

Figure 9.1. Prairie landscapes and key program areas.

wetland density and permanence were grouped, and then the boundaries of the landscapes were modified on the basis of other factors, such as landform, soil types, land use, local government boundaries, and the like. One such landscape from Saskatchewan, Conjuring Creek, is depicted in Figure 9.2. Each section (260 hectares) is

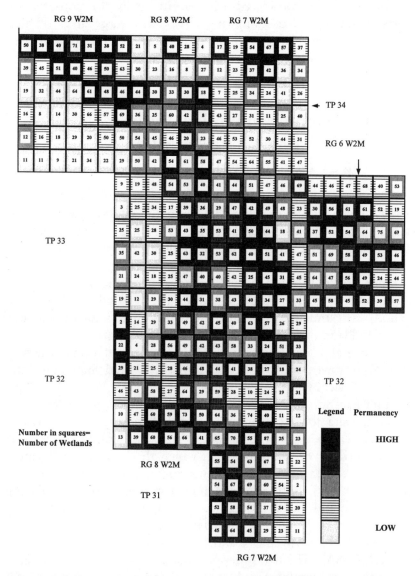

Figure 9.2. Conjuring Creek landscape sections ranked by wetland density and permanence.

ranked according to number of wetlands present (the numbers in each square are the number of wetlands) and also according to the number of hectares of permanent wetland. The higher the permanency rank, the greater the number of permanent wetland acres. Groups of sections with high wetland numbers and high permanency rankings were considered top-priority areas for conservation efforts. Landscapes like Conjuring Creek with large areas of high wetland density and large amounts of permanent wetlands were singled out as areas where protection and restoration efforts should be concentrated.

Program Planning within Landscapes: Baseline Landscape Characterization and Strategy Development

Once landscapes had been delineated on the basis of wetland characteristics, a wide variety of other data was gathered to allow characterization of the baseline, or pretreatment, habitats. Development of land management plans (habitat or implementation plans) that are designed to change the landscape from being generally unproductive and duck hostile to one that will improve recruitment to the extent that population goals will be attained was then carried out. Duck production was estimated through a modeling process for the baseline and treated landscape; the difference between the two is the estimated incremental production resulting from implementation of the program. Waterfowl population data from the United States Fish and Wildlife Service/Canadian Wildlife Service (USFWS/CWS) May Breeding Pair Surveys were the chief source of duck data used in the planning. Data for the Conjuring Creek Landscape are shown in Table 9.1. These are data from several segments along two transects in waterfowl survey Stratum 31, in which the landscape is located. Included are data on wetland basins containing water (ponds per square kilometer). Both are important data inputs to the modeling process. Although mallard data only are input into the model, outputs calculate total waterfowl population responses.

Wetland Habitat Inventory and land-cover/land-use data derived from LANDSAT Thematic Mapper image analysis for the targeted landscapes were supplemented by other data sources to provide baseline conditions (Table 9.2). These included Canada Land Inventory

Table 9.1. Breeding Mallards and Wetlands in Conjuring Creek Landscape, Saskatchewan, 1971–1990 (USFWS and CWS Breeding Waterfowl Surveys)

Year	Breeders/km²	Ponds/km²
1971	17.8	19.3
1972	12.0	17.5
1973	17.8	22.1
1974	4.9	30.4
1975	8.0	20.0
1976	9.5	14.0
1977	19.9	9.2
1978	7.1	12.3
1979	13.2	21.4
1980	11.4	8.9
1981	12.9	5.4
1982	10.2	19.9
1983	7.3	26.6
1984	9.1	14.6
1985	6.0	22.1
1986	7.7	15.0
1987	9.6	10.5
1988	7.7	17.4
1989	10.9	7.8
1990	7.6	14.2

Table 9.2. Land Use and Land Cover in Conjuring Creek Landscape Pre- and Postdevelopment (Area in Hectares)

Land Use/Cover	Predevelopment	Postdevelopment
Cropland	27,500.8	21,970.4
Fallow land	14,174.5	13,753.4
Direct seeded crops	850.6	4,737.2
Hayland	2,355.1	2,176.9
Pasture land	10,932.8	9,977.3
Underseed fallow	200.4	1,350.2
Delayed hay	0.0	323.9
Grazing system	0.0	793.5
New idle grass	0.0	161.9
Restored grassland (fenced)	0.0	40.5
Restored grassland (unfenced)	0.0	728.7
Open water	301.6	301.6
Deep marsh	273.3	273.3
Shallow marsh	4,600.8	4,600.8

and Forest Inventory data that provide broad land-cover classifications; Agriculture and Canada Census data on land use; and various other administrative databases related to crop insurance, municipal tax assessment, and the like.

The example landscape, Conjuring Creek, is an approximately 800-square-kilometer area on the eastern boundary of the Quill-Lakes-Touchwood Hills KPA, in the vicinity of Wadena, Saskatchewan. The majority of land in the landscape is Class 2 and Class 3 for agriculture (68 percent). About 32 percent is lower in agricultural capability (Class 5) on account of the presence of potholes and larger wetlands. The majority of the area is used for crop production (65 percent), with conventional tillage the norm. Only 32 percent of the land base is in grassland or similar cover (7 percent idle native and 25 percent pasture). Drainage of wetlands has been extensive in the region. Satellite inventory data indicate that there are 11,338 wetlands larger than 0.8 hectare that total about 7,300 hectares (1986 data). Pond densities range from 0.8 to 34 per square kilometer. Permanency was rated quite high.

Waterfowl populations have fluctuated from about sixteen pairs per square kilometer to over fifty-nine pairs per square kilometer over the last twenty years. At the time the plans were prepared, overall breeding populations for the landscape were estimated at about twelve pairs per square kilometer.

Other data on land tenure, rates for lease or purchase, soil types and productivity, size of farm units, and so forth were considered as well. A key component of the planning and data-gathering process was the assessment of landowner attitudes and preferences with respect to the various land management options ("land-use changes") that were proposed as part of the potential implementation plan ("landscape plan"). Determination of these factors was accomplished mainly through questionnaires and interviews. Prior to any detailed modeling and analysis, biologists and agrologists developed an overall strategy for the landscape based on all the factors outlined above (Figure 9.3). This represents a conceptual level of targeting within landscapes that attempted to integrate the biophysical characteristics of the landscape with the socioeconomic factors, which influence what protection and restoration activities are possible.

In the conceptual plan or strategy, areas with particularly high density and permanence of wetlands were targeted for intensive conservation action (that is, protection for exclusive wildlife use). In such

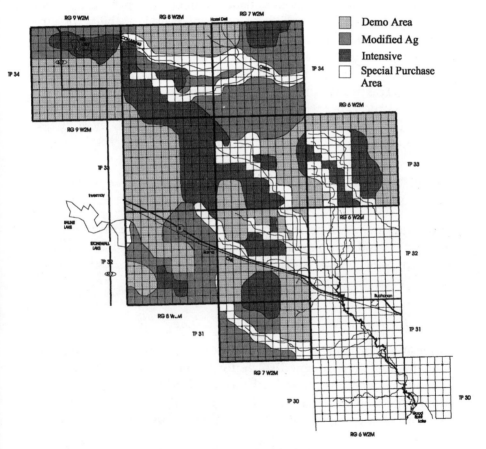

Figure 9.3. Delivery strategy map of Conjuring Creek landscape program.

areas, the wetlands were often associated with uplands that were heavily influenced by other land uses, mostly agriculture, but which were either marginal for agricultural production or had high percentages of existing grassy cover. Other areas, along streams or watersheds with the potential for or subject to high levels of wetland drainage, were specifically targeted for fee-simple acquisition to prevent further drainage activity. A comparison of Figures 9.2 and 9.3 shows that these two regions on the strategy map ("intensive" and "special purchase") correspond to the areas of Figure 9.2 that have the greatest density of wetlands and highest permanence rankings. Other areas that have good wetland numbers and permanence but are located in portions of the landscape where land capability for agricultural production is high were singled out as regions where attempts were to be made to influence the way in which agricultural production is carried out. It was hoped that at least partial restoration of such areas would make agricultural practices less detrimental to waterfowl and other wildlife production. Modification of agricultural practices and demonstration/extension programs were to be attempted ("modified ag" and "demo area" in Figure 9.3).

Program Planning within Landscapes: Options for Landscape Improvement

Major program components were grouped into "intensive uplands" (DNC), "modified agricultural use" (agriculture), "demonstration and extension," and "small wetland complexes and nesting structures."

Intensive Uplands

Intensive management options made no attempt to be agriculture friendly and were designed solely for wildlife production purposes. The idea was to protect and enhance the land involved in as short a time as possible. These techniques were specifically aimed at protecting and augmenting the size and distribution of grassland within the landscape.

Purchase and Conversion to Native DNC

Land units in the top 25 to 50 percent of sections that were ranked as high in wetland density and permanence and that had a diverse array of wetland sizes and types were targeted for purchase and conversion from cropland to native grass cover ("dense nesting cover," or "DNC"). It was also considered desirable to associate these purchases with larger blocks of existing native cover or to surround them with modified agricultural uses. This was also designed to restore connectivity in these fragmented landscapes. Wetlands that had been drained or highly modified were to be restored and wetlands in the surrounding area protected, so that broods produced as a result of the additional nesting cover would have better survival.

Native grasses, forbs, and shrubs were used because they are long lived, respond well to management procedures, and, as the diversity of species mix increases, provide habitat for the full range of upland nesting ducks and a wide range of other wildlife species. Twenty species of grasses (twelve cool-season and eight warm-season species), two legumes, four forbs, and five shrub or half-shrub species are in use in these restorations. Details about the use of native plant materials and techniques for revegetating with native grasses can be found in Wark et al. (1995). To provide greater diversity and longevity to these plantings than would be possible otherwise, an innovative technique known as "sculptured seeding" (Jacobsen et al. 1994) has been undertaken on an experimental basis.

Purchase or Lease of Native Parkland and Prairie

Native uplands in priority areas within landscapes may be purchased and managed purely as wildlife habitat. This option is of particular interest to many conservationists because it offers what many consider to be greater biodiversity and benefits to species other than waterfowl (Prescott et al. 1995). Limited areas of existing native pastures or hayland that have been carefully inspected to determine that they are in good shape and will require little or no renovation to be good wildlife cover are leased according to standard rates for such land in the jurisdiction involved.

Leasing of Cropland and Conversion to DNC

This is an option similar to purchase and conversion except that lease at the local going rate, instead of purchase, is the method of protection used. Leases are for a minimum of ten years, and rather than native cover, a mixture of introduced grasses and legumes is used to establish the nesting cover, primarily because the cost of using natives for the short term is not warranted.

A variation of this option is to lease to idle already-converted cropland. In some cases other programs, such as the Permanent Cover Program of the Prairie Farm Rehabilitation Administration (Agriculture Canada), have assisted farmers with conversion of marginal land to permanent forage. The Permanent Cover Program is similar to the Conservation Reserve Program (CRP) in the United States.

Modified Agricultural Use

"Modified agricultural use options" refer to a variety of land management procedures that may be delivered over large areas of the landscape and that are intended to result in development of benefits to wildlife without cessation of agricultural use. They were included in this project because they respond to the need to accommodate human uses within landscapes. They also offer potential to link ecosystem elements to one another (such as wetlands and grassland areas) and thus to restore connectivity and functionality.

Forage Delay Hay Cut

Payments to induce farmers to convert cropland to permanent cover were planned along with payments to delay the hay cut until after the middle of July, when most waterfowl nests will have hatched. Use of flushing bars in conjunction with, but also in lieu of, delayed haying operations can reduce or virtually eliminate hen mortality during harvest (Calverly and Sankowski 1995), so this option was also included. Delay hay cut was also planned for existing forage fields where landowners were willing. Payments are necessary because these practices are not agronomically sensible.

Inter-Pothole Seeding and Forage Establishment

Provision of specialized seeding equipment, saline-tolerant seed mixes, and custom seeding for restoration of pothole margins and inter-pothole areas was visualized as means to preserve wetlands and improve nesting cover around and between potholes. The primary goal was to maintain wetlands as pair space or brood salvage. These were also thought to offer the potential to restore diversity to the landscape and increase connectivity among the various ecosystem elements.

Pasture Management Systems

Pasture management systems were planned to replace continuous grazing regimes and thereby restrict or control the movement and number of cattle and the duration of grazing on existing pastures. Particular emphasis was placed on reducing the impact of cattle on wetland margins throughout the entire open water period, but particularly during the nesting period for waterfowl and other ground-nesting species. Management for residual cover and reduced disturbance during the nesting period is a basic principle. The results anticipated were drought proofing, better beef production, and increased waterfowl and other wildlife production. Better-quality grassland habitat was expected, as well as coexistence of wildlife and human uses.

Fallow Alternatives

Green-manuring, or underseed clover, is an alternative to conventional summer fallowing. It involves seeding of clover with a grain crop in year one, followed by fall harvest of the grain crop. The clover survives over the winter and is allowed to grow to form a cover crop or fallow substitute in year two. This is then either plowed down as a green manure or harvested for forage or seed. It is necessary to provide incentives to induce landowners to delay plowdown or harvest until after July 15 of the second year. This technique was included because if worked into a rotation, it provides the potential to replace, over large areas, conventional fallow, which is of virtually no value to the ecosystem.

Demonstration and Extension

Promotion of a variety of wildlife-friendly conservation farming practices, through demonstration and extension programs, was included. It was hoped that such indirect programs might eventually play an important role in modifying land-use practices over broader areas of the landscape than those that could be directly addressed by one-on-one interaction with landowners. Activities were to be concentrated in key program areas but not restricted to areas where landscapes had been delineated. Large groups of farmers and producer groups would be allowed to try new techniques without the expense of new equipment and the problems inherent in learning a new technology.

Functions for ecosystem management included improvement in over-all ecosystem health as well as potential for restoration of at least small measures of biodiversity and productivity. The potential to affect large proportions of landscapes otherwise unavailable for restoration would be the greatest benefit.

Promotion of the use of minimum tillage or under-cutter imple-ments to maintain standing stubble through the fallow year and into the next spring seeding period (stubble mulch fallow) and direct seeding of fall or spring crops into stubble was included as part of the conservation program. The techniques are expected to be self-sus-taining because they are clearly economically and agronomically sound and could affect large portions of landscapes if widely adopt-ed. In addition, cover that is provided by fall-seeded grain crops might very well provide grassland-like cover of significant benefit to ground-nesting birds.

Small Wetland Complexes and Nesting Structures

It was considered a fundamental requirement that small wetlands be secured as part of any protection and restoration program. These are often secured by the installation of nesting structures (baskets, rock islands) that are used primarily by mallards. In some areas they get 25 to 30 percent use with high nesting success, while in others very few are used. Even where nesting structures are not very productive, they serve a valuable purpose in providing for wetland protection and were therefore included in the planning and implementation of the pro-ject. Recent experience with nesting tunnels has shown that they are used to a greater extent and at very high rates of success, at least by mallards, making them attractive as both productive units and as tools to emphasize the value of the wetlands in which they are placed.

Land-Use Exchange

Land-use exchange is a way of influencing land use and protecting habitat in areas that have high wildlife potential but may not be obtained by purchase, lease, or other methods. They are the ultimate in responding to the need to accommodate human use and occupa-tion within ecosystems targeted for restoration. They were included as conservation program options for that reason. Their use can also serve the more biological requirements of ecosystem management, including restoration of diversity and increase in quality, quantity, and distribution of grassland habitat.

In a land-use exchange, the use of some land owned by conservation interests is managed in the way privately owned land is managed by the landowner. In effect, conservation interests use their land to bargain for changes in land use on the parcels owned privately but that are best suited for waterfowl and other wildlife production.

A typical exchange would have the project purchase a property and manage existing native uplands and any wetlands on the land as wildlife habitat. Local farmers would be allowed to cultivate or institute a managed grazing system on the good agricultural land on the property. In return, they would, for example, idle some existing pasture, allow restoration of grassland cover on some of their less-productive or difficult-to-cultivate acres, and allow restoration of wetlands on their properties.

Program Plan for a Landscape and Progress to Date

A computer planning tool (CPT) was developed to automate the complicated calculations required to estimate the costs and benefits of habitat securement, development, and management over a fifteen-year horizon. Use of the CPT allowed an iterative process of program design and cost-benefit calculation to optimize program plans so that implementation would be realistic, the costs reasonable, and the benefits in terms of incremental waterfowl production both realistic and near predefined objectives. Given the complexity and magnitude of the calculations, there would be no possibility of making more than one initial "best-guess" estimate of these parameters without the use of the CPT.

The CPT has two major modules: an economic module and the mallard productivity model (the latter adapted from Cowardin et al. 1988). A schematic flowchart (Figure 9.4) illustrates how the modules interact. The economic module allows the entry of land securement, development, and management plans, complete with acreage goals and associated direct and indirect costs. The overall costs of the landscape plan can then be projected over a set period of time (fifteen years is standard). The mallard productivity model allows the entry of baseline landscape characteristics (land cover, land use, wetland numbers, fluctuations over time, and waterfowl population data). The

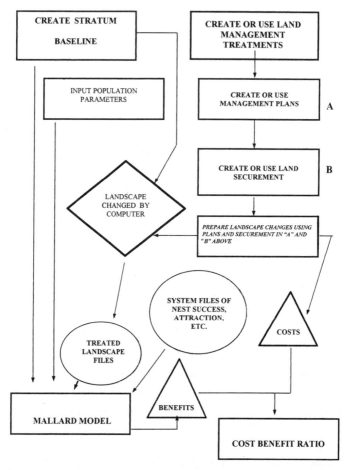

Figure 9.4. Computer planning tool relationships and flowchart. *Note*: Rectangles are planners' tasks, diamonds and circles are computer's tasks, and triangles are outputs.

two modules then interact, resulting in estimation of waterfowl benefits and calculation of cost/benefit ratios. More important, this modeling process allows estimation of the degree to which the habitat changes are likely to provide for population growth in the landscape. An example plan for Conjuring Creek Landscape in Saskatchewan is found in Table 9.3.

The plan is to treat 930 hectares with intensive upland options, 1,540 hectares with modified agricultural-use options, and 4,615 hectares with conservation farming demonstrations. Over 800 hectares of small wetlands will be protected, and about 100 hectares

Table 9.3. Summary of Program Plan for Conjuring Creek Landscape, Saskatchewan, Canada

Land-Use Option	Hectares Secured	Costs($)[1]	Ducks[2]
Uplands demonstrations			
Direct seed fall	729	10,500	4,368
Direct seed spring	3,887	128,896	7,466
SUBTOTAL DEMOS	4,615	139,403	11,834
Intensive uplands[3]			
Native unfenced DNC	243	372,672	9,813
Native fenced DNC	40	160,038	7,326
Introduced unfenced DNC	405	801,317	16,652
Idle existing native DNC	113	86,608	559
Lease hay to idle	81	89,690	3,271
Lease pasture to idle	49	27,119	240
SUBTOTAL INTENSIVE	931	1,537,444	37,861
Modified agricultural use			
Convert and delay hay	227	115,768	1,097
Delay hay existing	97	35,798	493
Grazing system	794	126,720	5,037
Underseed fallow	421	47,232	279
SUBTOTAL MODIFIED AGRICULTURE	1,539	325,518	6,906
TOTAL UPLAND HABITAT	7,085	2,002,365	56,601
Wetlands			
Nesting structures	70[4]	36,255	1,551
Small wetlands	914	17,425	n/a
TOTAL WETLAND HABITAT	914	53,680	1,551
TOTAL HABITAT SECURED	7,999	2,056,045	58,152

1. Costs for securement, enhancement, and management over 15 years.
2. Additional or incremental ducks added to the populations in landscape.
3. Meant to be secured and enhanced within first 5 years.
4. Number of nesting baskets or tunnels.

of wetlands will be restored. The expected waterfowl response to delivery and management of this habitat over a fifteen-year period is that breeding-pair numbers will increase to an average of about twenty-six pairs per square kilometer, or about 120 percent above the very low baseline. One of the model outputs is an estimate of the number of breeding ducks per square kilometer in the landscape over the fifteen-year modeling period (Figure 9.5). In this case, the trend is for a significant increase in breeding pairs as a result of treatment com-

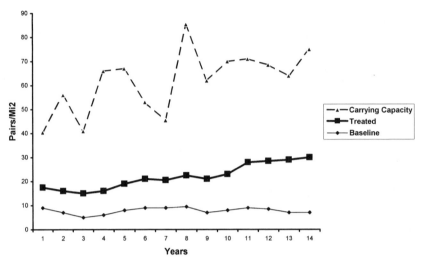

Figure 9.5. Treatment response—Conjuring Creek.

pared to the baseline condition, without exceeding the estimated carrying capacity.

The program is being monitored and evaluated through a series of specific, directed studies aimed at particular aspects of the various options being delivered, and by means of a major assessment (Prairie Habitat Joint Venture Assessment Study) of the overall response of waterfowl in both active delivery areas (treated landscapes) and controls (untreated). Results of these studies being carried out by the Ducks Unlimited Institute for Wetland and Waterfowl Research will not only tell us how well we are doing relative to waterfowl population goals, but also provide insights into how we can improve the program. Part of these evaluations and directed studies will also document the use of project areas by wildlife species other than waterfowl.

About 112 landscapes have been delineated across the three Prairie Provinces, with program plans now defined or in the process of being defined, following the process described above. Nearly 227,000 hectares of habitat had been protected or restored by September 1995 in sixty-seven major delivery areas as part of the conservation programs undertaken under NAWMP. An additional 81,000-plus hectares have been protected or restored in other landscapes where delivery is not a priority at this time, though quality projects are available and are secured when opportunities arise. The majority of these conservation efforts have occurred through Prairie CARE, Ducks Unlimited Canada's habitat conservation program.

Wildlife Responses

Recent studies have attempted to evaluate the overall value of NAWMP project areas to wildlife other than waterfowl or to document the use of specific land management projects by grassland passerines, small mammals, and herptiles. Prescott et al. (1995) ranked twenty-two habitat types, 75 percent of which were Prairie CARE sites, in the parkland region of Alberta relative to their overall value to wildlife. Wildlife species were ranked as to NAWMP priority, with migratory, sensitive, wetland-dependent, and endemic species having highest priority. Habitats with native grass and shrub species ranked higher than those with introduced species; and although large saline and fresh wetlands, idle deciduous uplands, and idle native parklands rated highest, idle native grassland and native-planted DNC ranked very close together (8 and 9, respectively, out of the 22). Given that about 30 to 35 percent of all land parcels purchased by the program consist of idle native grassland or idle native parkland, the Prairie CARE program is clearly focusing on priority habitats that serve a broad range of species and have considerable biodiversity values. In addition, Prescott et al. (1995) suggested that the program should focus on securing existing idle parkland because of its very high ranking and biodiversity values. This is in fact accomplished in the process of acquiring land for conversion to DNC, and by virtue of the fact that large tracts of intact native cover (often several sections) are targeted for acquisition.

Several authors have focused on the use by grassland birds of DNC or other managed covers, relative to cropland (controls) or other unmanaged cover (Dohl et al. 1995; de Sobrino and Arnold 1995; Hartley 1994). All indicate that significant benefits accrue to grassland birds when land is converted from cropland to managed covers, especially if native species are used in restoration. Both Dohl et al. (1995) and de Sobrino and Arnold (1995) worked in Manitoba and indicate that LeConte's sparrow (*Ammodramus leconteii*) and sedge wrens (*Cistothorus platensis*) were most abundant in DNC; both species were rare in the study areas prior to establishment of the cover. Savannah (*Passerculus sandwichensis*) and clay-colored sparrows (*Spizella pallida*), bobolinks (*Dolichonix oryzivorus*), common yellow throats (*Geothlypis trichas*), northern harrier (*Circus cyaneus*), and American bittern (*Botaurus lentiginosus*) also either preferred DNC or used it in high numbers relative to other cover types. Hartley (1994) reported similar findings. Sedge wrens were only found in DNC; bobolinks and common yellow throats were much more common in DNC than other covers; and several declining species—clay-colored sparrow, song sparrow (*Melospiza melodia*), red-winged blackbird (*Agelaius phoeniceus*), American goldfinch (*Carduelis tristis*), and bobolink—used DNC and existing native grasslands to similar degrees. Clearly, Prairie CARE components, like the intensive upland management options, have significant benefits to species other than ducks. Most important, some of the species apparently benefiting are declining in much the same way and likely for the same reasons causing declines in waterfowl populations.

Dale (1994, 1995) examined in a preliminary way the use of managed grazing systems by passerines. Endemic species such as Sprague's pipit (*Anthus spragueii*), western meadowlark (*Sturnella neglecta*), and Baird's sparrow (*Ammodramus bairdii*), all of which have declined in the grassland region, were found to be more abundant in deferred- or late-grazed paddocks in grazing systems than in continuously grazed pastures. In the 1995 report she noted that grasshopper sparrows (*Ammodramus savannarum*) were also more abundant in pasture systems. In that report she also indicated that general productivity seems higher in managed grazing systems, but also that more data are needed for confirmation.

Fisher and Roberts (1994) noted that grassland restoration and grazing systems in southern Alberta had significant benefits to wood frogs (*Rana sylvatica*), boreal chorus frogs (*Pseudacris triseriata*), tiger salamanders (*Ambystoma tigrinium*), plains garter snake (*Thamnophis radix*), and plains spadefoot toad (*Spea bombifrons*). Similarly, Skinner et al. (1995) concluded that planted cover, especially that which was three years old or more, had significant benefits to small mammals in parkland Alberta. Meadow (*Microtus pennsylvanicus*) and red-backed voles (*Clethrionomys gapperi*), deer mice (*Peromyscus maniculatus*), house mice (*Mus musculus*), jumping mice (*Zapus hudsonius*), prairie shrews (*Sorex haydeni*), pocket gophers (*Thomomys talpoides*), and ground squirrels (*Spermophilus tridecemlineatus*) were particularly abundant in such cover.

Relationship to Biodiversity and Ecosystem Management

Prairie CARE is a good example of ecosystem management in that it clearly integrates scientific knowledge within a complex sociopolitical and values framework toward the general goal of protecting native ecosystem integrity over the long term. This is a working definition of ecosystem management proposed by Grumbine (1994) that we think accurate and appropriate.

Key goals outlined by Grumbine (1994) focused on those most applicable to dealing with public lands and protecting "what's left." They include maintenance of native species and ecosystem types, maintenance of essential ecological processes, and accommodation of human use and occupancy. It is the last goal that receives the least attention from habitat managers but is one that is critical to ecosystem management in the Prairie Pothole Region. In such highly altered systems where human use and occupation is extensive, restoration of ecosystems and species populations is central to restoration of biological diversity and must solve the problems related to use by people.

Grumbine (1994) also noted that for programs to demonstrate sound ecosystem management, they must take a systems perspective and consider all levels of biodiversity. They must also work across administrative and political boundaries, defining ecological boundaries at appropriate scales to best maintain ecological integrity. Good science, monitoring and feedback loops, and application of adaptive resource management are required. Innovative and cooperative action is required, and humans and their values must be integrated into goal setting and implementation of programs. We believe that Prairie CARE conforms to all these requirements.

The Prairie CARE program is hierarchical and deals with ecosystems (landscapes), species (waterfowl primarily), and populations. Although political and administrative boundaries are factored into the planning and delivery of such programs, these are secondary to biological considerations. The primary factors that delineate areas to be used for program delivery are biophysical ones, such as wetland density and permanence and waterfowl breeding-pair populations.

The program is science based in that the best available information and technology are used to design and implement conservation activities. Data inputs to the mallard model and CPT are as up to date as possible. In addition, new information is acquired through research and evaluation (for example, the PHJV Assessment Project), with the goal of refining programs. We learn as we go. In other words, we practice adaptive resource management.

Biodiversity benefits occur through grassland restoration and long-term protection of prime blocks of "native," or relatively unmodified habitats. Likewise, conversion of currently cultivated land to native grass, forb, and shrub cover benefits a variety of wildlife species in addition to waterfowl. Options such as managed grazing systems, while providing for human use of the land base, also allow for improved wildlife production and benefits to biological diversity. Incentives provided to cattle producers induce them to convert from continuous, high-intensity grazing to managed systems, which improve cover quality and species diversity within pastures remaining in agricultural use. The key here is that there are agronomic and economic benefits to the landowner that will sustain improved habitat conditions. The improved habitat will provide significant benefits to biological diversity for the system involved.

Being innovative and working within the existing sociopolitical milieu are hallmarks of Prairie CARE and ecosystem management. Demonstration and extension programs and land-use exchange are prime examples of these principles in action. Our challenge in restoration of heavily impacted systems is to maintain existing partnerships, forge new ones, and, above all, to consider the people who own the land as real partners in the attainment of our goals. Only humans can create the socioeconomic and political adjustments needed to restore biological diversity and ecological integrity to the landscapes in which we coexist with wild species. Everyone involved needs to understand why conservation of biological diversity is important if we in the conservation community expect the general public to participate. If we are to cause people to invest in biodiversity, we need to clearly define what we want landscapes and ecosystems to look like in the future, temper that vision with the demands of a growing population, and then work with diverse interests to implement conservation programs.

Although the Prairie CARE programs focus on wetlands and waterfowl, similar approaches to other species groups will not only likely be successful in restoring populations, but will also be absolutely necessary. Small-scale, site-specific conservation activities and legislated protective measures will not be successful in restoring populations of wildlife species currently considered "at risk" or at the very least declining throughout the North American continent and indeed elsewhere. The cumulative effects of decades of habitat conversion and destruction, often on massive scales, can only be addressed by conservation programs that are similarly widespread, that are focused on large landscape units, and that integrate biological needs of wildlife with the socioeconomic factors responsible for the problems in the first place.

We need to appeal to the values and needs of all segments of society if we are to be successful in maintaining ecosystems and diversity of wildlife. Some will value biodiversity for its intrinsic values, but most will need to see more concrete benefits. We can "sell" ducks to duck hunters, songbirds to bird watchers, sustainability of agriculture to farmers, increased profits associated with improved land use to bankers, and contented constituents to politicians who shape land management policy and regulations. Many stakeholders must cooperate if ecosystem integrity and biodiversity are to be preserved in the PPR. The PPR is greatly modified but still remains a highly diverse and productive part of the North American continent. Prairie CARE and the NAWMP offer tremendous vehicles to return an ecological balance to the diverse landscapes throughout the region for the benefit of waterfowl and many other species.

If the integrated ecosystems approach taken in program planning and delivery for NAWMP were to be emulated by those aimed at endangered species, chances of real success would be greatly increased. There seems little to be gained by either protecting small pockets of habitat that harbor remnant populations of wildlife or undertaking expensive reintroduction efforts if the surrounding system is degraded, fragmented, and incapable of providing the necessary biological components to support those species.

A tremendous opportunity exists in the United States to integrate programs such as the Conservation Reserve Program (CRP) with programs aimed at managing ecosystems for maintenance of biological diversity and protection of wildlife populations. The CRP is large scale and widespread and is aimed primarily at managing agricultural production and soil and water erosion. If wildlife conservation programs were to focus on key areas within the regions affected by CRP and took an approach similar to that described here, substantial benefits to ecosystem integrity and wildlife conservation would result. Wildlife managers would be wise to take maximum advantage of the opportunity now available.

Chapter 10

Putting Diversity into Resource Conservation

Fred B. Samson
and Fritz L. Knopf

Theory based on the distribution and ecology of rare taxa is a reliable blueprint for increased action to conserve biological diversity (Bibby et al. 1993). The late twentieth century has brought exceptional concern worldwide for rare species and, at the same time, has seen many rare plants and animals brought close to extinction (Wilson 1994). In this chapter, we discuss the view that the best way to conserve biological diversity is to prioritize those areas with the greatest number of rare species, provide an initial priority list of major vegetation province types to conserve biological diversity in North America, and suggest how to put diversity into resource conservation. Although the usefulness of both single species and community ecology in conservation is debated, the theme of traditional conservation is and will remain around the single or group of species (Wilcove and Blair 1995).

Diversity: Where Is It?

Vertebrate and other taxa are limited in space and time at several scales (Brown 1995). The geographical distribution of single species is limited by physical factors (temperature regime, water availability, and so on) and biological factors (such as suitable habitat and the presence of competing species). Moreover, a single species is rarely found in all stages across a successional series and has a characteristic successional stage.

For groups of species, community ecology is the study of limits, attempting to explain the relative rarity of different species. Variation in use of space and time by species is huge, from plants that are nearly cosmopolitan in distribution (such as lambs quarter [*Chenopodium album*]) to plants that occupy small areas (such as Virginia round-leaf birch [*Betula uber*]). Plotting the frequency distribution of such a variety in geographic ranges for plants and animals would show two tails—one to the left, representing species with restricted geographic ranges, and one to the right, representing species that are cosmopolitan in distribution. Species that are not cosmopolitan in geographic distribution but are found in selected regions are termed "endemic" (Brown and Gipson 1983).

Wulff (1943) suggested more than 90 percent of the world's seed-plant flora are endemic on some scale. Endemism manifests itself at various taxonomic levels from variety to order. Cosmopolitan species account for a very small portion of the world's flora and are not evenly distributed across land areas of the world (Kruckeberg and Rabinowitz 1985). Some places, particularly islands and mountains, are rich in endemics while the arctic and boreal forest are relatively poor in them. Other areas—California, the European Alps, the Mediterranean region, Hawaii, alpine regions of central Africa, New Caledonia, the Cape Region of South Africa, and the Sino-Himalayan regions—are well-known centers of endemism.

As with flora, animals range in distribution, from the cosmopolitan peregrine falcon (*Falco peregrinus*) to the entire avian family Furnidae (ovenbirds), whose 56 genera and 214 species are restricted (endemic) to South and Central America (Brown and Gipson 1983). The 523 species of North American mammals are geographically very restricted (Pimm and Gittleman 1992), and species ranges tend to follow major habitat types (grasslands, wet forests, deserts, and so on), for these provinces are in part defined by their constituent species. Some regions, such as Australia, Southern Africa, Hawaii, Madagascar, and New Caledonia, are known centers of endemism. Thus, distantly related taxa of plants and animals tend to show patterns of endemism not only in the same ocean or island but also in the same localities within a continent.

The coincident distribution of endemics is referred to as "provincialism," and in the case of animals is influenced by both historic events (drifting continents, changing sea levels, and glaciation) and ecological processes (long-distance dispersal, biotic interactions, and extinction of small populations) (Brown and Gipson 1983). Distinguishing between historic events and ecological processes as a determinant influence on patterns in animal endemism is difficult. Debate continues with regard to the role refugia—generally high mountain ranges that have continuously supported the same vegetation type and associated fauna during glacial advances—may have played as the explanation of present-day patterns in species numbers and distributions (Huston 1994). The orientation of geographical barriers (mountain ranges and major vegetation provinces) in tropical and temperate regions of the Americas is a major influence on patterns in species distributions.

Studies further show that regions with high endemism are characterized by low productivity, especially where levels of soil nutrients and soil water are low (Huston 1994). In North America and elsewhere, high levels of plant endemism are associated with low levels of productivity, mineral nutrients, or light from high levels of precipitation. Patterns in vertebrate distribution in response to low productivity are more difficult to predict, but the relationship between body size, number of species, and available energy may provide a vehicle for making connections between the composition of animal communities and the net primary production that characterize the host systems (Brown 1995).

A large amount of ecological theory and concept derives from the study of the rare and endemic taxa. Darwin (1898) was first to identify that the quantity and quality of endemism differed among the major geographic, topographic, and vegetation types. Darwin noted that three primary factors—geographical area, ecological breadth, and isolation—describe the distribution of endemic taxa (Kruckeberg and Rabinowitz 1985). Today, these three factors have become central to the theory of island biogeography (MacArthur and Wilson 1967) and, in general, geographical ecology (MacArthur 1972) and conservation (Samson and Knopf 1982).

Diversity: What Is It?

An approach of growing interest in the conservation of biological diversity is to prioritize those areas with the greatest number of endemic species (Wilson 1994). This can only be accomplished by identifying the distinctiveness of individual areas and assessing their value at the global or continental scale (Ehrlich 1988).

Diversity is described by three parameters: the number of species in a specified area (alpha diversity), change or turnover in species across the landscape (beta diversity) (Samson and Knopf 1982), and hierarchical diversity (Mares 1992). Calculating alpha or beta diversity is not a trivial task, and the development of any approach to conserving diversity involves knowing both diversities (Pimm and Gittleman 1992). In addition, and long recognized (Samson and Knopf 1982), number of species per se is not the most important criterion for diversity conservation (Wilson 1994). Hypothetically, an area may contain large numbers of widespread, highly adaptable organisms with no immediate threat to their conservation. Furthermore, two sites similar in area may contain the identical number of species, yet one assemblage may consist of widespread, highly adaptable species of low conservation priority, the other of species found nowhere else.

A third issue in the conservation of biological diversity is that the area containing the most species may not conserve the greatest amount of genetic information (Vane-Wright et al. 1991), particularly if the set of species is closely related. Genetic diversity is maximized through conservation of distantly related taxa (Pielou 1975)—that is, conserving four species in four different orders versus four species within an order. This diversity is referred to as "hierarchical diversity" by Pielou; and it ensures that genetic diversity, not merely species diversity, is maximized and conserved.

One recent example is the conservation of Neotropical mammal diversity in South America (Mares 1992). Conservationists and scientists use the lowland rainforests in South America to rally support to conserve biological diversity. Nevertheless, one important question has yet to be answered: If only one macrohabitat could be chosen, which is most important to the conservation of neotropical mammalian biodiversity? There are six major vegetation provinces in South America: the drylands, Amazon lowlands, southern mesophytic (wet) forests, upland (dry) semideciduous forests, western montane forests, and Atlantic rainforests (Mares 1992). The two largest in area are the drylands and the Amazon lowland rainforests. Geographical ecology would predict that the species diversity of an area is set by area—that is, drylands ought to contain more species. For conservation, however, the question is which habitat (or habitats) has more unique species and high hierarchical diversity.

Examination of the distribution of neotropical mammals reveals a conclusion opposite to the research and conservation strategies employed to date (Mares 1992). Tropical rainforests are not as uncommonly diverse as frequently thought. The Amazon lowland rainforest supports fewer than 60 percent of the number of species found in other major habitats, and 68 percent of the species found in the Amazon lowland rainforest are shared with other major habitats. Nearly all genera (greater than 95 percent) found in the Amazon lowland rainforest are shared with other major habitats. In contrast, the non-Amazon habitats have more species than the Amazon lowland rainforest (509 versus 434) and share fewer species (less than 39 percent) and genera (less than 60 percent) with other major habitats. This pattern is even more pronounced in the comparison of endemism in the six major habitats.

More endemic species (53 percent) and genera (44 percent) use macrohabitats other than the Amazon lowland rainforest. About 2.5 times (336:138) as many endemic species are found in major habitats other than the Amazon lowland rainforest, and about 6.5 times (65:10) as many endemic genera are found in non-Amazon habitats. Overall, the Amazon lowland rainforest supports fewer species, fewer endemic species, fewer genera, fewer endemic genera, fewer families, and fewer endemic families than the drylands (Mares 1992:978). Thus, an optimal conservation strategy for South American neotropical mammals begins by increasing the number of protected dryland areas, recognizing that the size of an area and the number of taxa (particularly endemics) are related, and accepting that establishment of further Amazon lowland reserves is misguided as far as neotropical mammals are concerned.

There is no single list of priorities in the conservation of biological diversity in North America (Samson and Knopf 1994). In fact, the worldwide priority areas for conservation established by the International Council for Bird Preservation includes but one for North America, the Sierra Madre area in California. Moreover, there is only one complete continental analysis in North America that identifies possible conservation areas for species likely to disappear in the foreseeable future (Dobson et al. 1997). The Dobson et al. (1997) algorithm maximized the number of endangered species (animal and plant) while minimizing the area required to do so. Their results imply that the aggregated distributions of endangered species are the product of centers of endemism and anthropogenic activities (urbanization, agriculture, and so on) and suggest that birds (and perhaps arthropods) act as important indicators for the presence of other endangered taxa.

The extent to which the protection of habitats for endangered species may or may not conserve biological diversity is unknown. An alternate approach to conserving biological diversity is to prioritize those areas with the greatest number of endemic species (Wilson 1994) and to consider hierarchical diversity (Mares 1992).

The sixteen major vegetation provinces in North America are six dry, open provinces (the Arctic, Paulose grassland, California grassland, Mojave and Sonora deserts, Great Plains, and Chihuahuan desert); four dry forest provinces (the Boreal, Intermountain, California, and Rocky Mountain); and six wet-forest provinces (the Pacific, Mixed Deciduous, Appalachian, Piedmont, Coastal, and Tropical) (Barbour and Billings 1993). (See Table 10.1.) An analysis of bird distribution to identify areas of endemism is an essential and accepted approach to identifying priority areas for conserving biological diversity (Bibby et al. 1993), and a number of recent studies suggest this to be the case (Collar and Stuart 1988, among others). The list of bird species in North America covered in Table 10.1 was compiled from maps in standard references (Poole and Gill 1998). The major habitat(s) associated with each species was organized by the sixteen major vegetation provinces to allow a simple overview. Other information displayed in Table 10.1 includes the level of endemicity to a major vegetation province by genera, family, and order.

The analysis of the distribution of North American bird taxa—18 orders, 54 families, 258 genera, and 575 species—by major habitat type is revealing in terms of conservation priorities (Table 10.1). The six dry, open habitat types support more species (476) and more endemic species (27) than either the dry-forest (16 and 375, respectively) or wet-forest (13 and 328, respectively) habitat types. In particular, the prairie of the central Great Plains contributes to the high levels in species numbers and endemism characteristic of the dry, open habitat types. The Boreal and Rocky Mountain forests contribute in a meaningful way to the number of species and level of endemism evident in the dry-forest formation. The tropical habitat type is important within the four major wet-forest formations, although species numbers and level of endemism are low. Few avian families or genera are restricted in distribution to one vegetation formation in the United States and Canada. The dry, open and dry-for-

Table 10.1. Major Vegetation Provinces in North America and North of Mexico (Barbour and Billings 1993) and the Distribution of Shared and Unique Endemic Bird Taxa

Vegetation Formation	Order (No.)		Family (No.)		Genera (No.)		Species (No.)	
	SHARED	UNIQUE	SHARED	UNIQUE	SHARED	UNIQUE	SHARED	UNIQUE
Dry, open								
Arctic	17	0	31	0	130	0	210	7
California grassland	14	0	36	0	107	0	124	4
Chihuahuan	15	0	34	0	82	0	113	1
Great Plains	18	0	46	1	184	1	292	12
Mojave/ Sonora	16	0	40	0	117	0	171	3
Paulose	16	0	39	0	115	0	168	0
Dry forest								
Boreal	17	0	41	0	130	0	217	7
California	14	0	42	0	113	0	150	3
Inter- mountain	16	0	43	0	140	0	190	0
Rocky Mountain	18	0	47	1	187	0	306	6
Wet forest								
Appalachian	15	0	34	0	82	0	113	1
Coastal	16	0	39	0	105	0	129	0
Mixed deciduous	15	0	38	0	131	0	198	2
Pacific	15	0	41	0	125	0	187	3
Piedmont	15	0	33	0	92	0	112	0
Tropical	15	0	37	0	83	1	100	5

est major habitat types support more orders, families, genera, and species than the wet habitat types. Thus, in the United States and Canada, to conserve the greatest amount of bird diversity, one would choose the dry, open forest and dry, open vegetation formations and place less priority on wet forest.

The grasslands of the Great Plains that extend south from Canada into Mexico and west from the Rocky Mountains into Wisconsin and Indiana are the largest vegetation province in North America (Samson and Knopf 1994). Thus, it is not surprising that area and number of species and number of endemic species are closely associated. It is difficult here, as in South America (Mares 1992) or elsewhere (Bibby et al. 1993), to suggest that hierarchical diversity (at least above the species level) is a significant issue in establishing priority areas for conservation in North America as far as birds are concerned. Nevertheless, if one major habitat is important to the conservation of North American birds, it must fall within the dry, open habitat type as the priority in conserving diversity.

Over 90 percent of biota listed as threatened or endangered in the United States are narrow endemic taxa (Samson and Knopf, unpublished data). Dobson et al. (1997) suggested that a large portion of the endangered species in the United States can be conserved in a few key areas, especially in open and the dry habitats of California, the intermountain West, and the Mojave and Sonora deserts (Dobson et al. 1997:551). Given the need to conserve a complement of species beyond those at immediate risk, there is nonetheless considerable similarity in major habitat types in North America that support both endangered species and a high level of avian endemism. Dry, open habitats tend to be more important to the conservation of endangered and endemic birds, and a general pattern emerges that less-productive systems do in fact support higher levels of biological diversity.

Diversity: Science and Management

Most conservationists and scientists recognize that (1) not all species are equally susceptible to extinction, (2) endemic taxa are prone to extinction, (3) certain areas are exceptionally rich in endemics, (4) scientific information must be the foundation to conserve biological diversity, and (5) there is a need to incorporate such information into proactive protection plans (Pulliam and Babbitt 1997).

The Center for Conservation Biology at Stanford University has developed an initial tool for analyzing human-related threats to global biodiversity (Sisk et al. 1994). It is based on the geographical patterns in species distribution, number of endemic species, species habitat use, habitat loss, and trends in human population growth. The approach builds on that of Myers (1988) in his regional analysis of regional conservation through the identification of regions of exceptionally high plant species diversity and endemism and the ranking of countries and regions within countries on the basis of the highest number of endemic birds thought to be at risk by the International Council for Bird Preservation (Bibby et al. 1993). The purpose was to draw on country-specific data on diversity, endemism, and threats to native habitats to prioritize conservation activities.

Sisk et al. (1994) suggested that continental or regional analyses are a next logical step in the formulation of conservation strategies, and one study (Dinerstein and Wikramanayake 1993) illustrates such an attempt. An impediment to implementing continental and regional strategies is lack of data, and few, if any, continental or regional strategies to prioritize conservation activities are available. While these few analyses are to be applauded, most conservationists and scientists with conservation interests are often frustrated by the lack of data to describe all populations constituting the flora and fauna of a region or even an area. Managers, either public or private, acquire methods and knowledge from consultants, continuing education, the literature, and, above all, their own experience (Davis and Botkin 1994). As a consequence, agency- and organizational-dependent terminology and methods divide the real world in natural resource management. Data—words, sounds, images, and numbers—are the realm of science. Information—arranging data into meaningful patterns—is the bridge between science and management. Knowledge—information, skill, and vision—is management. Merging the three key elements—data, information, and knowledge—leads to "information sludge" (Mintzberg 1994) and contributes to today's ineffectiveness in acquiring the data, information, and knowledge essential to establishing global, continental, and regional conservation strategies.

Angermeier and Karr (1994) ascribed much of the loss of biological diversity in the United States to an ineffective policy that emphasizes piecemeal conservation of elements rather than the comprehensive protection of the integrity of native species assemblages. Understanding the organization and management policy of governmental agencies (Knopf 1992) and conservation organizations (Howard and Magretta 1995) is as fundamental and important as improved understanding of what and where biological diversity is. Journalist David Osborne was recruited by Vice President Al Gore to reinvent the business of government, "replacing large, centralized, command-and-control bureaucracies with a very different model: decentralized, entrepreneurial organizations that are driven by competition and accountability to customers for results they deliver" (Posner et al. 1994:136). The notion that biological diversity can be managed from lower to higher has limited utility. One government working in cooperation with nongovernmental organizations is required to agree on which organizational levels and diversity elements should be protected and the methods to acquire needed data, information, and knowledge.

Shifting our everyday thinking is required to put diversity into resource conservation. One of the most important challenges is understanding how systems characterized by low productivity and low species number contribute disproportionately to biological diversity and that all species are not created equal. Critical steps toward managing for biological diversity include understanding alpha, beta, and hierarchical diversity; identifying the distinctiveness of individual areas on the basis of the three forms of diversity and assessing their value at the global or continental scale; the acquisition of consistent and credible data, information, and knowledge; and a clear and central policy on the unique value of diversity by government and nongovernmental organizations. In addition, a clear message from the growing literature on the conservation of biological diversity is the need to maintain the native ecological processes that shape and maintain the diversity characteristic to a landscape (Samson and Knopf 1996). Such shifts are, however, not likely to favor the conservation of biological diversity unless people recognize the inherent value of rare biological elements and native ecological processes at all spatial and organization levels. This is the most important issue in putting diversity into resource management.

Opportunities
and Challenges

Part 3 investigates an array of issues that can deter managers from attempting to achieve conservation of biological diversity. Chapters 11 to 13 bridge these difficulties by providing a range of alternatives for managers to consider and apply.

Chapter 11 addresses a common concern: how best to initiate a strategy for conserving biological diversity. Nudds points out that uncertainty is prevalent, credible advice is sadly lacking, and conflicting views are often the order of the day. Unfortunately, ever since the Rio Conference established the formula "Inventory, Monitor, and Assess" (IMA), both scientific and ethical controversies have swirled about what, exactly, to inventory and monitor and what, exactly, to assess to best conserve biodiversity even just at the species level. Nudds offers a solution, encouraging managers to use adaptive management. This strategy attempts to reduce uncertainty and contribute to improved knowledge. Such an approach requires, however, that the IMA triad be reordered, to "Assess, Monitor, and [lastly] Inventory" (AMI).

Next, Chapter 12 provides the perspective that biodiversity conservation must be based on sound science and questions whether that is currently occurring in all possible situations. Trauger notes that current knowledge of the composition, distribution, abundance, and life cycles of most species of plants and animals is incomplete, insufficient, unreliable, or nonexistent. Contemporary managers are also confronted with additional levels of complexity related to varying degrees of understanding about interactions of species and ecosystems. The author notes that traditional species-oriented management schemes may have unintended consequences, and that ecosystem-oriented management initiatives may fail in the face of inadequate or fragmentary information on the structure, function, and dynamics of biotic communities and ecological systems. Nevertheless, resource managers must make decisions and manage based on the best biological information currently available. Whereas some level of management must exist to meet agency responsibilities, the author believes that more research is needed to conserve biological diversity.

Chapter 13 examines the delicate management situations needed to cope with exotic species and restore locally extirpated species. Campa and Hanaburgh alert managers to the need for special consideration in these scenarios, which often result from misdirected management practices of past generations. The chapter discusses considerations, challenges, and opportunities for reducing the threats of exotic species to biodiversity and for reducing chances of additional introductions. Examples illustrating what biologists have learned about exotic species provided a background for management approaches that minimize their impacts on biodiversity.

Chapter 11

Adaptive Management and the Conservation of Biodiversity

Thomas D. Nudds

Until relatively recently, the fields of natural resources (wildlife, forestry, and fisheries) management and ecology developed on the basis of fundamentally different questions about nature (Nudds 1979; Nudds and Clark 1993; Nudds 1998). Resource managers principally concerned themselves with maintaining sustained yields; ecologists were more concerned about why populations fluctuate (Lancia et al. 1996:438). However, even before the Rio Summit (Johnson 1993) coaxed a closer relationship between "basic" and "applied" ecology, there were calls for fundamental changes in the natural resources management professions (for example, wildlife, Capen 1989; fisheries, Walters 1992; forestry, Society of American Foresters 1993) and some scientific societies (such as the Ecological Society of America; Lubchenco et al. 1991) to better integrate basic and applied ecology to learn how humans might best extract resources while conserving biodiversity and the productive capacity of ecosystems.

Fundamental to these changes was, and still is, the growing realization among resource scientists and managers that nature is capricious and that a great deal of uncertainty underpins theory about the dynamics of populations and communities (Ludwig et al. 1993). Thus, a great deal of uncertainty accompanies management models based on ecological theory. To proactively deal with uncertainty while managing resources requires a fundamental shift in the way management is conducted (Lancia et al. 1993; Lancia et al. 1996). In effect, because knowledge about nature is imperfect, managers must literally "learn while doing"—what Macnab (1983) referred to as "management by experiment" and what is more broadly referred to as "adaptive resource management" (ARM) (Walters 1986).

I begin this chapter by dissecting its title, defining each term— "adaptive management," "conservation," and "biodiversity." My point here is to argue that adaptive management, though developed (at least implicitly) to assist in, and most often referred to in the context of, conservation of populations of harvested species, is also relevant to matters regarding the conservation of biodiversity at the scale of landscapes and communities. Against that backdrop I discuss the roles of inventories and monitoring in adaptive management. Finally, I discuss some examples concerning the effects of forest management on birds. The examples range from strictly manipulative management experiments, at reasonably large spatial scales (about 72 square kilometers), which are conducted as part of ongoing forest management, to examples that integrate existing databases to evaluate the effects of forest management at huge spatial scales (greater than 235,000 square kilometers).

Adaptive Resource Management

The concept of adaptive resource management (ARM) can be traced back twenty years to a group of ecologists, which included Holling, Walters, Ludwig, and Hilborn, with interests in resources management at the University of British Columbia. Recently, biologists and managers of the U.S. Fish and Wildlife Service and the Biological Resources Division of the U.S. Geological Survey successfully implemented, continentwide, an adaptive management program for renewable harvests of mallard ducks (see, for example, Williams and Johnson 1995; Williams et al. 1996; Johnson et al. 1997). Similar programs for northern pintails and American black ducks are proposed. These applications of adaptive management feature populations of individual species as the foci for research and management attention. More broadly, ARM is done whenever the dual goals of achieving management objectives and gaining reliable knowledge are to be accomplished simultaneously (Lancia et al. 1993, 1996). Thus, the principles of ARM can be applied to the conservation of communities or landscapes (Schmiegelow and Hannon 1993; Clark and Diamond 1993; Conroy and Noon 1996) as well as to populations.

The goal of researchers is usually to obtain better knowledge about how and why biological systems behave as we observe them to, whereas the goals of managers more often involve some desired response by the system to management intervention, such as sustained or restored biological diversity in the face of resource extraction. From the perspective of managers, these goals converge because progress toward desired outcomes increases when uncertainty about future management decisions is reduced through learning (Lancia et al. 1996). From the perspective of scientists, ARM offers the practical advantage of meeting the needs of managers so that factors of interest can be manipulated at appropriate scales of space and time, with adequate replication and statistical power to yield reliable inferences (Macnab 1983).

Macnab (1983) coined the phrase "management by experiment." Experiments are essential to learning by means of the scientific method—namely, that from a series of observations and assumptions hypotheses about why the observations take the form they do are developed, predictions that arise from the hypotheses are subjected to "experiments" (of a variety of forms) (Hurlburt 1984), and the results of the experiments are evaluated against the predictions of the hypothesis (Figure 11.1). If the predictions are confirmed, then the hypothesis is considered a reasonable explanation for the observations, until such time as better hypotheses are invoked and tested. However, if the predictions of the hypothesis are not borne out by experiments, then the hypothesis itself is rejected or modified and/or the nature of the observations and the assumptions that gave rise to it are reevaluated.

In analogous fashion, ARM is a means to evaluate the effects of various resource management policies on populations and communities. Lancia et al. (1993) offered that *experiment* is to *hypothesis* as *management* is to *policy* (Figure 11.1). Sometimes the criticism is leveled that the validity of policies—as products of values and knowledge—cannot be subjected to scientific analysis; thus, policies cannot be treated as hypotheses. Such criticism fundamentally misses the point of adaptive management. Certainly, the validity of some policies cannot be judged by recourse to scientific scrutiny. The Vatican, for instance, has a policy of not ordaining women as priests, for whatever reason. Nevertheless, the validity of policies with regard to the management of natural resources can often be scrutinized scientifically because they are at least underpinned by ecological (as opposed to moral, ethical, or spiritual) observations and assumptions. These

ADAPTIVE RESOURCE MANAGEMENT
Policy as hypothesis, Management by experiment

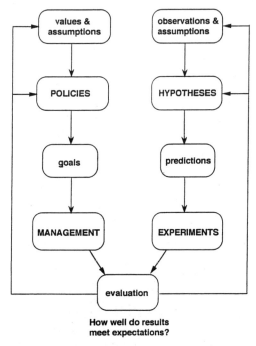

Figure 11.1. A flowchart emphasizing analogous components of scientific analysis and policy evaluation.

are testable. For example, a policy to allow spring hunts for black bear could be instituted because there are those who value hunting from economic and social perspectives. However, behind such policies are assumptions regarding the potential to conduct harvests while sustaining populations—for example, whether harvest mortality and natural mortality are compensatory. So, in the same manner that hypotheses can be revised or rejected because the assumptions that underpin them are invalid, so too can resources management policies be revised or rejected if the opinions and assumptions that underlie them turn out, on closer inspection, to be invalid (Figure 11.1, Clark 1992). Science is simply a way of knowing (Moore 1985); attitudes and values can evolve in the light of improved knowledge. In the next section I develop a policy statement for the conservation of biodiversity that is, in essence, a scientific "null hypothesis."

Conservation

Oliver et al. (1995) described conservation, like beauty, as in the eye of the beholder. *Funk and Wagnalls Standard Desk Dictionary* defines "conserve" this way: "To keep from loss, decay, depletion, maintain," and "conservation" as "the act of keeping or protecting from loss or injury. The preservation of natural resources, as forest, fisheries, etc., for economic and recreational use."

Importantly, from a scientific perspective, these definitions imply that to know whether a system manipulated for resource extraction by humans exists in a conserved state requires a comparison with a reference state (Arcese and Sinclair 1997). In the absence of any such comparison, it is not possible to judge whether the goal of conservation is met. In scientific terms, the reference state is the "control" for the suspected effects of anthropic change. Historical data alone can be used to document change, but by themselves provide limited ability to infer the cause of change. With no control, there can be no conclusion.

Similarly, because "ecological integrity" is also in the eye of the beholder, Karr (1991) developed a definition for the term that incorporated the idea that to know ecological integrity when it is observed requires a comparison with a reference system deemed to possess it. Thus, the onus lies with those responsible for managing resources while conserving ecological integrity and biodiversity (and who would use science to defend any claim that they are successful at it) to demonstrate that where anthropic change takes place, the system does not differ, in ecologically significant ways, from a similar system not subjected to anthropic change.

For instance, a system manipulated by forest cutting would be deemed to retain "ecological integrity" when it has "species composition, diversity and functional organization that is representative of natural habitats [that is, without forest cutting] within a geographic region" (U.S. Environmental Protection Agency 1993, based on Karr 1981). In this way, a policy that expresses a desire to harvest trees while conserving ecological integrity, or biodiversity, can be restated formally as a scientific "null hypothesis" (Figure 11.1). In effect, then, the goal (Figure 11.1) is to manage forests in such a way as to not enable rejection of the null hypothesis that species composition, diversity, and functional organization of the ecological communities are the same in areas where forests are harvested and where they are left alone. In this context, it is apparent why large, intact protected areas, to serve as baselines or controls (see Stohlgren et al. 1995; Arcese and Sinclair 1997) for the effects of anthropic change outside of them, are important.

Two caveats are required. First, it is obvious that at some arbitrarily small scale, some differences can always be found between manipulated and reference areas. Cavity-nesting animals will not occupy the precise spot where a tree with cavities once stood after the tree is cut down. From the perspective of an individual cavity-nesting animal, the removal of such a tree may perhaps be an ecologically significant event, but from the perspective of a population of such animals, or the entire assemblage of cavity-nesting species, it is potentially trivial. A goal of adaptive management is to identify thresholds where resource extraction exceeds the capacity of an ecosystem to "absorb" its effects without significant change to ecosystem integrity.

Second, criticism has been raised that, because humans are part of ecosystems, it is inappropriate to consider ecosystems without the effects of people as reference areas. It is not possible in the space available here to delve into the messy philosophical issue about whether humans can be considered "natural" (Hunter 1996; Haila 1997), but neither is it necessary. From a scientific standpoint, the issue is this: If the objective is to learn about the effects of anthropic change on the integrity of ecological systems, then a reference area without the presumed anthropic effect present is a *requisite*. Anything less is scientifically indefensible. Whether it is ever really possible to find reference areas really removed from all human influence is quite another matter (Arcese and Sinclair 1997), but that needn't deter us from at least constructing "thought experiments" that include "no-human-influence" reference areas for the purposes of evaluating the null hypothesis elaborated above. As an example, there are no such things as pure black bodies or frictionless surfaces, yet such reference points serve theoretical and empirical physicists well for studying the effects of changes in radiant energy and surface properties of objects. In practice, ecological reference sites will be chosen where the presumed effects of some anthropic change are at least minimized or some particular type of anthropic change is absent. Thus, national parks might serve as ecological baseline controls for the effects of anthropic change outside them, even though they may be used by humans for a variety of low-impact activities (Arcese and Sinclair 1997; Nudds and McLaughlin 1997).

Biodiversity

Diversity is the state of variety among objects in a collection. Although there have been accumulated numerous definitions of "biodiversity" (see Chapter 1), it simply refers to the state of variety among objects in collections of biological material. In practice, "biodiversity" often refers narrowly to the variety of species, or species richness, present at a particular place and time. Because the category "species" occupies an intermediate position in each of two intersecting hierarchies for classification of biological systems—one taxonomic and the other functional (Figure 11.2)—species richness is often, at least implicitly, assumed to reflect the variety of the components higher or lower than species in each of the hierarchies and used as a surrogate measure of biodiversity. Further, and perhaps even more

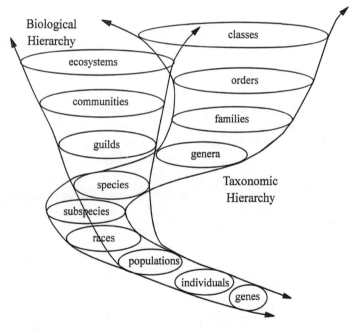

Figure 11.2. Two intersecting hierarchies of biological organization, one "taxonomic," the other "functional." To some extent both are arbitrary classifications for biological material, the products of human minds in need of systems for organizing information. Importantly, "species" occur at an intersection of both hierarchies. This is why, though it is currently debated whether taxonomic and functional diversity are synonymous, species diversity is nevertheless often used as a surrogate for biodiversity at other levels up and down each hierarchy (after Nudds 1997).

contentious (Baskin 1994; Moffat 1996), biodiversity indexed by species richness is often assumed to relate to function, or ecological processes, in ecosystems, although the empirical evidence is ambiguous (Johnson et al. 1996; Grime 1997).

Thus, there has developed considerable interest in conducting inventories of species. Indeed, the Convention on Biological Diversity (Johnson 1993:85) states, in Article 7, that

> Each contracting party shall . . . [i]dentify components of biological diversity . . . [m]onitor, through sampling and other techniques, the components of biological diversity . . . [and] [i]dentify . . . activities which have or are likely to have significant adverse impacts on the conservation of biological diversity, and monitor their effects.

In short, scientists and managers are to inventory, monitor, and assess (IMA). For some, this has been taken, explicitly or implicitly, also to be the *order* in which these activities should be undertaken. However, two significant problems are immediately confronted under that assumption.

First, it is not clear (1) that we can define well the entities (species) that are to be inventoried and monitored, especially many plants and undescribed invertebrate taxa; (2) if they could be, whether or not it is possible to achieve a complete inventory of species in a dynamic and fluctuating world, and if not, how complete would be "complete enough"; and (3) whether the cost of complete inventories is justified and the monies better spent on other conservation initiatives (Nudds 1997). As an alternative to "complete," systematically and taxonomically "correct" inventories are inventories using methods of "parataxonomy," or indicator species; for the latter, the question can still be posed: Indicators of *what*, exactly? Clearly, in the absence of knowledge about what is to be assessed (that is, what the question is), it is difficult to design inventory or monitoring programs that will serve the needs of assessment.

Second, related to the first problem, gross inefficiencies can result when programs to inventory and monitor biodiversity are undertaken without some clear understanding about what questions (hypotheses) need to be addressed *before* data are collected. Inefficiencies that arise from hypothesis-free data collection programs fall into three categories: (1) the data that are collected turn out to be useless in assessing any presumed anthropic effects (and therefore end up gathering dust in filing cabinets); (2) data already resident in filing cabinets and perfectly adequate for the task of assessing effects of anthropic change on biodiversity are overlooked in the rush to inventory and monitor (that is, to gather new data); and (3) scientists who collect data and interpret them after the fact are more vulnerable to committing errors of retroductive logic—that is, confusing speculation with explanation about patterns in data (Romesburg 1981). When this happens, time, effort, and money may be directed to remedial measures for "nonproblems" or "wrong" problems. Glenn and Nudds (1989) and Schmiegelow (1992) illustrated this problem for research that had been conducted on the effects of habitat fragmentation and insularization on the diversity of mammals and passerine birds, respectively, in terrestrial habitat isolates.

Answers to questions about what affects biodiversity and how best to conserve it will no more emerge serendipitously from species' inventories than will monkeys at a typewriter produce the Lord's Prayer. Thus, the protocol implied by the order of tasks in Article 7 (Johnson 1993:85) might be reversed (AMI): Scientists and managers need to ask what factors are hypothesized to affect change in biodiversity and therefore need to be *assessed*. Then, what needs to be *monitored* to undertake assessment becomes readily apparent. Finally, because monitoring can be considered nothing more than a series of sequential *inventories*, what needs to be inventoried should also be readily apparent.

The importance of evaluating hypotheses about cause-and-effect relationships for designing cost-effective programs for inventory and monitoring and deciding among options for redressing environmental problems is illustrated well by several examples in which adaptive management techniques have been used, in both manipulative and mensurative designs of management experiments, to evaluate the effects of forest management on bird diversity.

Effects of Forest Management on Bird Diversity

Recent research and management activity in the area of forestry-wildlife interactions in Alberta, Missouri, and Ontario are significant in that, although they differ in details and scale, each represents attempts to learn about the effects of forest harvest practices on wildlife concomitant with commercial forestry. The emphasis in

Missouri (Kurzejeski et al. 1993) and in some projects in northern Alberta (Stelfox 1995) has been to test hypotheses about the effects of forestry by establishing, thus far, sampling regimes for *selected* wildlife response variables (that is, an inventory) as a first step to monitoring changes in those response variables under various forest management practices. The next two sections discuss adaptive management projects in Alberta and Ontario that run from a highly controlled and replicated manipulative experiment conducted by a forest company as part of commercial operations (Schmiegelow and Hannon 1993; Schmiegelow et al. 1997) to a large-scale mensurative experiment using existing satellite imagery and bird distribution data to evaluate hypotheses about the effects of forest management (Gurd 1996). The final section describes how an adaptive management program was conceived for Ontario in response to terms and conditions that resulted from a Class Environmental Assessment of Timber Management for the province. Bird diversity and abundance figure prominently in these examples because birds are sensitive indicators of forest condition (Morrison 1986) and are more easily and completely inventoried than other taxa (Nudds et al. 1996), and also because bird species richness correlates with the species richness of other taxa (McLaughlin 1997). Thus, bird diversity may be a rough, surrogate measure of biodiversity.

Effects of Forest Cutting on Birds in Alberta

Alberta Pacific Forest Industries was granted timber rights to over 61,000 square kilometers of mixed-wood boreal forest in north-central Alberta, and agreed in a statement (reminiscent of changes in U.S. policy and law in the early 1970s that caused a virtual revolution in forest management practices) to "maintain viable populations of all resident wildlife species with good geographic distribution throughout the forest management area." Schmiegelow and Hannon (1993) modified an existing, commercial clear-cut harvesting plan to study the response of bird communities before and after experimental forest fragmentation. In this manner, they could scientifically evaluate progress toward the policy goal while improving knowledge about the effects of habitat fragmentation on bird diversity by conducting a manipulative experiment. Following a priori power analysis to determine the most cost efficient experimental design, the company cut trees such that there were six complete replicates of four size classes of fragments from 1 to 100 hectares; half in each category was connected by riparian corridors and half was not. An equal number of control plots was established in a large contiguous block left unharvested.

Certain operating ground rules for forestry in Alberta (for instance, regarding the need to leave riparian buffers) and the adequacy of existing tree inventories for planning harvest and roads placed some constraints on the experimental design. But even constrained experimental designs are apt to lead to the accumulation of reliable knowledge more quickly than are the alternatives (Nichols 1991; Arcese and Sinclair 1997). With this design, the researchers can ask whether forest fragments have fewer species than control areas the same size; whether there is nonrandom loss of species from fragments; and whether there are some spatial and temporal thresholds of landscape change at which these effects are not observed. Importantly, the company learns how to harvest in ways that should mitigate these effects and thereby meet the goal of maintaining biodiversity while extracting resources.

Effects of Clear-cutting and Fire on Birds in Ontario

In boreal forests, high-intensity crown fires are a prevalent and natural stand-replacement disturbance. Before the present era of fire suppression, fires are estimated to have occurred at intervals of 20 to 135 years, to average over 700,000 hectares, and to have produced a dynamic patchwork of habitats of varying ages, each with many different species. At a regional level, then, such disturbance is thought to enhance boreal biodiversity, birds included. However, these observations led to a contentious idea: namely, that clear-cuts, especially large ones, merely substitute for the natural disturbance regime. While it is trivially obvious, in terms of habitat structure and spatial arrangement of successional habitat patches, that fires and clear-cuts produce dissimilar *patterns* on the landscape, it is not so clear that, above some threshold, there is no difference between the effects of the two kinds of disturbance on the ecological *processes* that sustain bird diversity.

Gurd (1996) tested the null hypothesis that bird diversity and abundance were the same on landscapes created by fire as those created by clear-cutting. He conducted a "mensurative" experiment using existing LANDSAT forest cover data overlaid by data about the

size and distribution of clear-cuts and fires and data about bird distributions from the *Ontario Breeding Bird Atlas*. From over 235,000
square kilometers of northern Ontario, he selected 394 100-square-
kilometer plots (354 with a history of fire disturbance only and 40
with a history of mostly clear-cutting) for which there were reliable
data about the occurrence of eighty-seven species of birds. Logistic
regression was used to evaluate the null hypothesis that the probability of occurrence of each species was related in the same way to landscape configuration regardless of the type of disturbance. Gurd
(1996) found that for only six bird species did their probability of
occurrence not differ with the type of disturbance, that the probability of occurrence differed between fire and clear-cutting for forty-
eight species, and that the results were ambiguous for thirty-three
species. However, which species were affected and which were not
did not vary with any ecological differences among the species, such
as their sensitivity to forest block size, migration strategy, whether
they were cavity nesters, or the seral stage with which they were associated. Thus, there was little evidence that fires and clear-cuts are sub-
stitutable disturbances, even at very large spatial scales; the hypothesis that they are similar ecological disturbances with respect to their
effects on birds is false.

Forestry Policy and Adaptive Management in Ontario

While large, mensurative experiments can be conducted at scales
appropriate to the level of the phenomena under study, they are
sometimes nevertheless constrained by the quality of existing data
(Gurd 1996). Ideally, at very large scales, adaptive management programs, *sensu stricto*, should be conducted to gain the dual advantage
of obtaining the most reliable knowledge while managing.

Ontario is uniquely poised to implement adaptive forest management on a grand scale for two reasons. First, the province is legally
required to implement the terms and conditions set out in the Class
Environmental Assessment into Timber Management Board
Decision (Ontario Environmental Assessment Board 1994).
Specifically, Terms and Conditions 80 and 81 state, respectively, that
"the Ontario Ministry of Natural Resources (OMNR) shall undertake
long-term *scientific* studies to assess . . . the effects of current timber
management practices on wildlife habitat . . ." and OMNR ". . . will
monitor population trends of terrestrial vertebrate species" [empha-

sis added]. Second, a key policy goal of Ontario's strategic forest policy (Ontario Forest Policy Panel 1993) and law (Revised Statutes of Ontario 1994) is that forest management practices, including methods of harvest, will emulate natural disturbance.

Thus, the conditions for adaptive forest management are present (Figure 11.1). There is a policy statement (and a legal requirement) that can be recast as a scientific null hypothesis: Forest management has effects not different from those of natural disturbances. The hypothesis makes an explicit prediction to test, or a policy goal: There is no difference in ecological integrity (species composition, diversity, and functional organization) between places where timber management is practiced and where it is not. Further, the adaptive management framework leads to a series of alternate hypotheses with different implications for further development of policy and management guidelines. For example, if the null hypothesis was not rejected, at some spatial scale of analysis, then there wouldn't be any necessity for new guidelines at that scale. However, if at some scale the null hypothesis was rejected, then to meet the goal set by policy and law to manage forests to emulate natural disturbance would require some new or modified guidelines to bring the managed system within the limits of variability in the natural system. The limits of variability for the natural system can be measured concomitantly on adjacent reference (or control) areas of sufficient size.

Summary and Conclusions

It is becoming increasingly obvious that the management of renewable resources is plagued by uncertainty (Ludwig et al. 1993), especially with the requirement to ensure the conservation of biodiversity and the productive capacity of ecosystems while extracting resources. The uncertainty arises from two sources: nature itself is often quite variable and unpredictable, and our concepts and models about how and why natural systems behave as they do are incomplete and, in some sense, always will be.

To proactively deal with uncertainty requires more than merely arguing that the United Nations insists on adoption of the "precautionary principle"—namely, that absence of proof of alleged anthropic degradation of biodiversity shouldn't be used to stall remedial measures. By that route, uncertainty only persists about whether anthropic activities actually have the alleged effects, and any learning taken from the experience is based on the least reliable scientific knowledge. Further, in such cases where resource extraction is halted or regulated differently, some might argue that human welfare is unnecessarily compromised. In that sense the "precautionary principle" is a double-edged sword. Presently, because it is invoked to guard against a "type II error"—that is, wrongly accepting a false null hypothesis about anthropic effects on ecosystem integrity—it is biased to "type I error" about anthropic effects. That is, it is premised on the idea that we should assume that the null hypothesis of no anthropic effects is false, in case it is, and until it can be proven that it isn't false. Thus, the "precautionary principle" imposes a bias against ecosystem manipulations like resource harvests, which arguably poses a different kind of moral and ethical dilemma, especially in developing nations. Adaptive management, as a means to gain the most reliable knowledge about the response of ecosystems to anthropic change (learn while doing), is a means to minimize both types of error.

Earlier I alluded to and addressed two common criticisms of adaptive management: (1) it is an inappropriate intrusion of science into the realm of feelings, values, and policies, partly the product thereof, and (2) it is inappropriate to use (or perhaps impossible to find) reference areas removed from the influence of humans; therefore, good experiments about the effects of humans on ecosystems are technically unfeasible. These kinds of arguments have been addressed by others (Nudds and Morrison 1991; Arcese and Sinclair 1997). To those criticisms have been added that (1) adaptive management merely presents an opportunity for scientists to fiddle while Rome burns, and (2) experiments, insofar as they are also manipulations, carry a risk of negative effect on species or ecosystems; so it is better to do nothing than to fool with it. In this version of the "precautionary principle," there is a biodiversity "crisis" and an urgent need to act, even without reliable knowledge about the effects of alleged remedial activities, especially because we have only one world and can't afford mistakes.

Einstein is reputed to have said, when asked what he would do if he had an hour to save the world, that he would spend the first fifty-five minutes thinking about the problem. He knew that, if problem solvers identify the wrong problem in the first five minutes, it may not really matter what remedial measures are applied in the next fifty-five (unless by serendipitous good fortune somehow the right problem gets corrected for the wrong reasons—hardly a risk worth taking with only one world!).

Neither is adaptive management advocated by wild, Frankensteinish scientists bent on reliable scientific knowledge at all costs. For example, no one would separate the last of a breeding pair of an endangered species to subject one individual to some treatment while the other is left as a control, but this has been, incredibly, a criticism of adaptive management. Even so, adaptive management is advocated for rare and endangered species programs, especially when programs can be designed to test models of minimally viable populations (Boyce 1993) or involve reintroductions that can and should be done experimentally (Sarrazin and Barbault 1996; Armstrong et al. 1994).

Further, even early attempts at experimental harvest management of single species' populations invoked "fail-safe" measures. Waterfowl management programs designed to evaluate the merits of stabilized hunting regulations (Brace et al. 1987), for example, set population sizes for various species, so that if any dropped below "fail-safe" levels, the experiment would be reevaluated. The current adaptive management scheme for waterfowl populations also hedges against excessive harvests if populations get very small as a result of environmental conditions (Johnson et al. 1997).

Finally, adaptive management, because it requires large, intact natural areas to serve as reference sites with which to evaluate the effects of anthropic change in other landscapes (Arcese and Sinclair 1997), can help to further the cause of setting aside large protected natural areas distributed widely across the variety of ecosystems that humans manipulate (Nudds and McLaughlin 1997).

Chapter 12

Can We Manage for Biological Diversity in the Absence of Scientific Certainty?

David L. Trauger

Biological diversity is complex beyond understanding and valuable beyond measure (Ryan 1992): it is central to the productivity and sustainability of the Earth's ecosystems (Christensen et al. 1996). But on a global scale, biological diversity is threatened by increasing human populations and accelerating anthropogenic interventions in ecosystems (Wolf 1987; Wilson 1988; McNeely et al. 1990; Wilson 1992). Habitats are being degraded and destroyed, and species are declining and disappearing on an unprecedented scale (Ehrlich and Wilson 1991; Ryan 1992).

Conservation of biological diversity depends on sound scientific information about underlying ecological processes (Trauger and Hall 1992; National Research Council 1993); yet our current knowledge of the composition, distribution, abundance, and life histories of most plants and animals is incomplete, insufficient, unreliable, or nonexistent, confounding attempts at management. Managers are confronted with additional levels of complexity related to varying degrees of knowledge and understanding about species and ecosystems (Miller et al. 1985; Trauger and Hall 1992; Sampson and Knopf 1993).

Biological systems are highly complex, inherently dynamic, and exceedingly variable in response to natural processes and anthropogenic impacts. Consequently, traditional species-oriented management approaches may result in unintended consequences, and ecosystem-oriented management initiatives may fail because information on

the composition, structure, function, and dynamics of biotic com-
munities and ecological systems is inadequate or fragmented
(Committee on Environment and Natural Resources 1994; U.S.
Government General Accounting Office 1994).

The Resource Manager's Dilemma

The central issue of this chapter—and for conservation of resources—
is whether it is possible to manage for biological diversity in the
absence of scientific certainty. For resource managers, this is more
than a rhetorical question. This is the reality of everyday experience
for most land managers, who must make decisions and operate programs on the basis of the best biological information currently available and, simultaneously, further agency or corporate management
objectives. But rarely is expert knowledge sufficient for the analysis,
prediction, and management of any given resource management situation (Ehrenfeld 1991) According to Weston (1992), science has
played only a minor role in the conservation of biological diversity.

The answer to the question, "Can we manage for biological diversity in the absence of scientific certainty?," can be yes, no, or maybe,
depending on one's perspective. Although these responses appear to
be contradictory, convincing arguments can be made in support of
each. Many wildlife and wildland managers could emphatically
answer yes, because they have to manage every day. Given our present state of biological and ecological knowledge, they have no alternative but to do the best job they can with the scientific information
available to them (Meffe and Carroll 1994).

But these same managers could justifiably answer no as well,
because they always have some level of knowledge about the biological resources and ecological systems for which they have management responsibility. Generally, managers are motivated by a strong
sense of professionalism to seek out the latest information and to
consult with recognized experts before implementing their management schemes. They want to do the right things and to avoid doing
the wrong things in managing the resources under their stewardship.
However, scientists may not be aware of the knowledge and experience that managers bring to their work, and vice versa (Schonewald-
Cox 1994).

Recently, a new breed of resource manager has emerged, one who
would answer "maybe" in response to the management question.

These pragmatic managers recognize that they are dealing with substantive scientific uncertainties (Walters 1986). In response, "adaptive management" has gained proponents and practitioners. Adaptive management is a way of dealing with uncertainty in the management of natural resources (Lanica et al. 1993; Chapter 11).

Adaptive Management

Natural resource managers work in a highly competitive world of declining operational budgets and mounting societal pressures for more or less intense exploitation of resources. Professionally, managers face substantial scientific uncertainty but increasing responsibility for the conservation of biological diversity. Many ecosystems appear to be changing faster than scientists are gaining an understanding of them. Meanwhile, agency funding for biological and ecological research has generally been declining in recent years.

Adaptive management offers a resource management paradigm that allows managers to learn something about the species or systems that they are managing while they are managing (Lanica et al. 1993). Adaptive management has been used primarily in evaluating harvest management for fishery, wildlife, and timber resources (Walters 1986), but adaptive management strategies are also being applied in situations other than harvest management (see Chapter 11). By designing resource management manipulations as "experiments," managers are able to learn about the potentials of natural populations through knowledge gained in the process of managing the resource. Many managers view adaptive management as the preferred alternative to a reliance on conservative management strategies, which may be wasteful of renewable resources and greater organizational investments in basic research that may pay off only in the long term.

Adaptive management is no panacea; potential pitfalls lurk in such approaches. Biologists have observed that ecosystems are not only more complex than we think, but also that they are more complex than we can think. In addition to lack of control over the primary physical, chemical, and ecological processes, managers lack control over social, economic, and political parameters affecting resource management options (Ehrenfeld 1991). Realistic management goals may be difficult to identify because the determination of what constitutes biological knowledge and ecological criteria is so elusive. A more common problem in adaptive management, however, is not that some form of science is involved, but rather that the "best available science" seems seldom used. Conventional wisdom and anecdotal science seem often to inform the design of the management scheme.

Furthermore, some management responsibilities do not lend themselves to adaptive strategies. For example, adaptive management is particularly risky in management of endangered species, where every individual is important for biological diversity at several levels—genetic, population, species, and ecosystem. Likewise, granting development permits in accordance with adaptive management principles may unduly risk irreversible impacts on fish and wildlife habitats or irretrievable damage to other natural resources. In this regard, the recent widespread implementation of Habitat Conservation Plans for protection of endangered species seems fraught with peril, given the absence of scientific research on their efficacy.

In any case, managers probably would avoid using adaptive management strategies in implementing habitat protection programs because they must usually take whatever areas they can get rather than being able to experiment with alternative reserve locations and designs. A population declining because of unknown biological or chemical agents—such as irrigation drainwater in the Central Valley of California (Ohlendorf et al. 1990)—is another situation in which adaptive management would be a questionable approach because the priority objective is to find the cause and source of the contamination. Managers may use adaptive management in the restoration of damaged or destroyed ecosystems, but such approaches may be inefficient if the desired biological composition and ecological function endpoints are unknown. Nor would this be the preferred management response to control of invasive nonindigenous or exotic plants and animals where the damage to biological diversity may be rapid and extensive. In all of these situations, adaptive management approaches may be inappropriate or ineffective at the very least and may result in unintended consequences.

The Case for Research

Basic, long-term ecological research represents an investment in our future (National Science and Technology Council 1995). Sound science is critical to avoiding counterproductive management programs (Trauger and Hall 1992; Schonewald-Cox 1994). In the absence of adequate science, resource managers and policy makers will have to make decisions and take action in the face of higher degrees of uncertainty and risk (Committee on Environment and Natural Resources 1994). Greater reliance on scientific information is the only rational way for resource management agencies to avoid irreversible processes, irreparable damage, and unintended consequences affecting biological diversity. Without sound experimental design to ensure that the methods of data collection are adequate to demonstrate effects, conservation strategies may be seriously misguided and conceptually flawed (Mares 1992; Trauger and Hall 1992). Furthermore, resource managers and biologists, as well as other conservation advocates, will not have credibility in challenging those who would advance development projects that further jeopardize biological diversity. Scientific information, based on properly designed and conducted studies, is necessary to prevent advocates for biological diversity being dismissed as extreme environmentalists. Cutting-edge research must lead the way to new insights and innovative approaches to conserving biological diversity (Soulé and Kohm 1989; Lubchenco et al. 1991).

To do a good job of advocating and managing for biological diversity, we need long-term, large-scale, multidisciplinary studies (Committee on Environment and Natural Resources 1994). Such research must increasingly be problem-driven rather than discipline-oriented science. A major research need is the development of adequate models to evaluate the effects of management practices on biological diversity (Van Home and Wiens 1991). In implementing ecosystem management schemes, natural resources managers and biologists are developing a complex of ecological models that attempts to portray the various relationships between ecosystem components as they vary across the landscape and through time. Scientists are most comfortable when predictions and conclusions are grounded in scientifically demonstrated relationships, but some of the assumptions on which these models are based remain unsubstantiated by research. Currently, there are conceptual, informational, and technological deficiencies affecting our ability to perform in this area.

The New Policy Context

Conservation of biological diversity is emerging as a central focus of public interest in the environment and is a driving goal for a host of agencies, organizations, and corporations (Chadwick 1990; Salwasser 1991; Lubchenco et al. 1991; Haufler 1995). Ecosystem management is rapidly becoming a policy initiative in response to the continuing loss of biological diversity at all levels and across many administrative jurisdictions and land ownerships (Grumbine 1994).

Clearly, managing natural resources on the basis of sound scientific information is better than managing without it, but "good science" does not make "good policy" (Meffe and Carroll 1994). Indeed, the link between science and policy is as yet poorly understood and developed (Gordon and Lyons 1997). Another reality is that scientists often focus on the *lack* of information as a basis for new studies, whereas managers focus on *available* information as a basis for decision making (Interagency Ecosystem Management Task Force 1995). While we are making progress on several fronts toward the goal of improving our scientific knowledge of ecological systems and biological diversity (Ehrlich and Wilson 1991), we have significant public policy and budget challenges ahead.

The imminent and immense pressures of rapid human population growth during the next decade will set in motion a mad scramble to develop housing and infrastructure (Morrisette 1992; Committee on Environment and Natural Resources 1994). As agricultural and economic development intensifies to meet escalating needs and rising expectations in both developed and developing nations, so too will ecological strains. The accompanying competition for both renewable and nonrenewable resources will cause even greater stresses on biological diversity and ecological systems locally, regionally, nationally, and globally than is occurring today.

Protection of biological diversity in rapidly urbanizing areas appears to be a particularly urgent emerging challenge. A priority research focus should be on areas that are urbanizing and where loss of biological diversity will be occurring dramatically. Urban planners need to know what options are available to them to make better decisions on land uses and development patterns. And the reality is that we do not have much time remaining to act in many areas if we hope to make a difference in conserving biological diversity.

Although there is much to do and little time, the good news is that there are many opportunities for all to contribute toward the goal of conservation of biological diversity (Trauger and Hall 1992). New partnerships are emerging among agencies, organizations, corporations, groups, and individuals (Trauger et al. 1995; Haufler 1995). Programmatic consensus about a shared vision of desired outcomes is needed among all stakeholders before management changes can be implemented. The goal is worthy of our best collective efforts.

The days of squandering scarce fiscal resources on unscientific, trial-and-error management schemes should be behind us. But as public funding gets tighter and tighter, every special interest will be increasingly intent on ensuring their place in state, provincial, and federal budgets (Trauger et al. 1995). Research dollars will be increasingly scarce, and every scientific discipline will mount highly competitive campaigns for the remaining budgetary scraps. At present rates of fiscal outlay for debt service and other entitlements in the U.S. budget, there will be *no* discretionary funding in any federal agency by 2012 (Lamm 1997). Surely, the budget crunch will hit many programs and agencies within the next five years. Strauss and Howe (1997) predicted a major national crisis, a chain reaction of emergencies possibly triggered by an unyielding impasse over the federal budget and the ensuing financial collapse of the United States around 2005.

Summary and Conclusions

Conservation of biological diversity is emerging as a major societal goal. Because natural processes are so complex, can we wait until science finds answers before we begin managing for biological diversity? A definitive answer to this question may be impossible to find, but the need for one is gaining in importance and immediacy. To the research scientist the idea of proceeding without adequate information is unthinkable, but to the resource manager it is a fact of everyday life, now and for the foreseeable future. Meanwhile, a public policy consensus, that greater reliance on credible science will be crucial to efforts to protect and manage biological diversity, is emerging.

While the goal of conserving biological diversity may still be achievable, major changes in our management priorities and research agendas will be necessary. Even though management for biological diversity may be infeasible or inefficient without a sound scientific basis, some level of ongoing management must be done to meet agency responsibilities and societal expectations. In the face of current constraints, we may have to shift personnel and programs away from traditional discipline-focused approaches and place emphasis on multi-disciplinary problem-focused efforts.

In the end, many key management decisions affecting biological diversity will essentially be gambles, no matter how they are justified and presented, but sound scientific information can substantially reduce the risks. Scientists must help managers make better decisions through greater synthesis and use of existing scientific information. More research is also needed to reduce the uncertainty in efforts to manage for biological diversity. An improved scientific basis for conservation of biological diversity is possible over time. All effort should be made to achieve this objective as rapidly as possible.

Chapter 13

A Management Challenge Now and in the Future: What to Do with Exotic Species

Henry Campa III and
Christine Hanaburgh

For decades, natural resource managers have dealt with the challenges of managing different types of species. Managing species presents a special challenge for natural resource managers, one requiring knowledge of their habitat requirements and population characteristics. Some native species, such as the bald eagle (*Haliaeetus leucocephalus*), Eastern wild turkey (*Meleagris gallopavo silvestris*), and giant Canada goose (*Branta canadensis*), have required extensive management to increase their numbers. As a result of those efforts, their populations have rebounded to levels that will ensure their continued existence within their native ranges. But other human activities have either intentionally or accidentally caused species to shift their locations so that local populations have expanded their ranges, and species have been introduced to new areas distinct from their historic ranges. This chapter discusses the challenges that exotic species present and opportunities for reducing their threats to biodiversity and the chances of additional introductions. It presents several examples to illustrate what biologists have learned about exotic species, and it proposes management approaches for minimizing the impacts of these species on biodiversity.

What Are "Exotics"?

In 1993 the Office of Technology Assessment estimated that at least 4,500 species (from plant pathogens to vertebrates) foreign to the United States have established sustainable populations here. Furthermore, other species native to specific regions of the United States have been introduced in areas beyond their historic ranges. Such species that become established artificially outside of their historic ranges (that is, the ranges species evolve in) through some method other than natural dispersal are referred to as "alien," "exotic," "introduced," "nonindigenous," or "invaders." For example, species that evolved in Europe, such as the house sparrow (*Passer domesticus*), are considered to be exotics to North America and native to Europe.

In this chapter, the term "exotic species" is defined as species that occur outside of their native spatial boundaries (the ranges they evolved in) because of deliberate or accidental activities by humans. Species that undergo natural range expansion from dispersal that has not been facilitated by humans will not be considered exotics here, although the cause of range expansion may be debatable for some species.

Numerous exotic species have been introduced because they are useful to industry and other human activities. Examples (and their respective uses) are iris (*Iris* spp.) (horticulture), the European honey bee (*Apis mellifera*) (apiculture), the soybean (*Glycine max*) (cattle feed), the African clawed frog (*Xenopus laevis*) (biomedical research), rust fungus (*Puccinia chondrillina*) (a biological control) (Schoulties 1991), rainbow trout (*Oncorhynchus mykiss*) (sportfishing), and ring-necked pheasant (*Phasianus colchicus*) and chukar (*Alectoris chukar*) (hunting).

Some species introduced deliberately have affected ecosystems in ways that were not predicted; others have been accidentally transported or have invaded susceptible areas weakened by disturbance. Moulton and Sanderson (1997) have stated that species from nearly all taxonomic groups have been introduced somewhere, many causing deleterious impacts on ecosystems. Exotic species may threaten biological diversity through direct competition with native species, direct effects on native species through predation, parasitism and disease transmission, and/or the alteration of the composition, structure, and ecological processes within ecosystems. Exotic species that affect specific ecosystem characteristics are potentially the biggest threats to biodiversity, but their effects are often underestimated and unnoticed until they eventually evoke a significant reduction of native species or the extinction of a unique ecosystem. By then it is often too late to mitigate the ecological damage that has been done. Species that have caused such problems are numerous and include the zebra mussel (*Dreissena polymorpha*) (Nalepa and Schloesser 1993), Austrian pine (*Pinus nigra*) (Leege 1997), red imported fire ant (*Solenopsis invicta*) (Pedersen et al. 1996), cheatgrass (*Bromus tectorum*), melaleuca (*Melaleuca quinquinervia*), and purple loosestrife (*Lythrum salicaria*).

Keys to Successful Establishment of Exotic Species

For exotic species to become successful (that is, develop self-sustaining populations), adequate environmental conditions must be present. These may include mechanisms for them to be transported into new geographic areas. If an exotic is going to be an exotic, it first has to be transported outside of its historic range. The greater the number of mechanisms and the ease by which a species can be transported, the greater chance it will become successful in some other geographic area.

For species to become established outside of their historical ranges, the new territory must provide hospitable conditions and climate that maintain their survival. A species transported to a subtropical or tropical climate will likely have a greater chance of becoming established than a species transported to a region with temperate or Arctic conditions. Subtropical or tropical climates insulate species from harsh winter conditions (Moulton and Sanderson 1997). At least 106 species of vertebrates have become established in Hawaii and 72 in Florida (Moulton and Sanderson 1997), most likely due in part to their mild weather conditions and the availability of resources for food and cover.

Native species within their historic ranges evolve under unique, dynamic disturbance regimes, both spatially and temporally. When these dynamic regimes are altered, new habitat conditions may develop favoring the infusion of exotic species (Moulton and Sanderson 1977) and hindering the continued existence of native species. Hanaburgh (1995) observed this type of phenomena on grasslands in southwestern lower Michigan. Remnant native grasslands throughout her study area lacked an abundance of many of the native plant species that historically dominated these types of grasslands. She concluded that the absence of fire on the native grasslands (for almost thirty years) could have contributed to the decline in the uniqueness of these areas. In essence, an alteration in the disturbance regime beyond the system's historical range of variability may have created ecological conditions that were no longer well suited for the native plant species.

Exotic species that become established in a region can have numerous, direct negative effects on ecosystem processes and compositional attributes, while positive effects are limited. Occasionally, however, exotic species have played roles in maintaining some characteristics of ecosystems where similar native species have been extirpated.

Negative Effects of Exotic Species on Natural Ecosystems

Although exotic species have been propogated in many areas with the intention of improving the quality of wildlife habitat and/or increasing the diversity of wildlife species for recreational pursuits, there are many situations where exotics have threatened the biodiversity of an area. Negative effects of exotic species may occur when a planned introduction affects an ecosystem in ways that were not predicted, or when an exotic species is unintentionally introduced in an area. Two main mechanisms of unintentional species introductions are through the transportation of a species beyond its native range and through habitat disturbances that increase the susceptibility of an ecosystem to invasion. Four specific negative impacts of exotic species that can lead to a deterioration of biodiversity are predation, direct competition, ecosystem alteration, and disease or parasite transmission.

Predation by an exotic species on native species may be the most immediately observable impact that exotics can have in ecosystems. Predation may ultimately eliminate native species from an area. Native prey species located on isolated "islands," either oceanic or terrestrial, may perhaps be especially vulnerable. More specifically, those prey species that lack coevolutionary experience with an exotic predator may be especially at risk of extinction.

In California streams, the eggs and larvae of the California newt are preyed upon by two exotic species, a species of crayfish (*Procambarus clarkii*) and the mosquitofish (*Gambusia* spp.). California newt eggs are generally protected from native predators by a gelatinous capsule that surrounds the eggs and by a neurotoxin within the body of the newt embryos. However, these defenses did not evolve in the presence of the exotic predator species, and they are ineffective defenses against crayfish and mosquitofish (Gamradt and Katz 1996).

Savidge (1987) documented perhaps one of the most well-known examples of an exotic species potentially causing changes in the relative abundance and composition of a community through predation. Following the introduction of the brown tree snake (*Boiga irregularis*) to Guam after World War II, some bird species disappeared by the 1960s in southern Guam. Savidge (1987) noted that by the mid-1980s, many of the species occurring in northern Guam had also disappeared, presumably owing to the spread of the brown tree snake. Rodda and Fritts (1992) also attributed the brown tree snake as causing the disappearance of the spotted-belly gecko from Guam's forest.

Direct competition with native species is another damaging effect of exotics. Exotic species often outcompete native species for food or breeding sites, limiting the growth of the native population, inhibiting its reproduction, or displacing it from its habitat. Two prominent examples are the intentional releases of the European starling (*Sturnus vulgaris*) and house sparrow in the United States in the 1800s, which have been responsible for competition with native songbirds and displaced some native bird species (Ehrenfeld 1970). Worldwide, nearly 20 percent of the cases of avian nest displacement documented in the scientific literature have been caused by introduced house sparrows or starlings (Lindell 1996). For example, in rural areas of the United States, house sparrows aggressively invade and take over the nests of the eastern bluebird (*Sialia sialis*), a native cavity nester whose numbers have declined in some parts of the country (Sauer and Droege 1990; Pogue and Schnell 1994).

Such incidences of direct competition are not limited to terrestrial ecosystems. McKnight and Hepp (1995) documented how an exotic species in the United States, the grass carp (*Ctenopharyngodon idella*), can compete with native wildlife species for the same forage and have a direct negative effect on the composition of a plant community. Grass carp are used occasionally in some regions of the United States to control the spread of aquatic plants to facilitate recreational activities such as boating and swimming. One problem associated with this practice of biological control is that when the grass carp is stocked at relatively high rates it may forage heavily on submerged vegetation (Bain 1993), especially in closed systems. Often the carp prefers native plant species (Pine and Anderson 1991) rather than the exotic species present. The use of native aquatic vegetation by the grass carp is therefore problematic for wildlife species that are also dependent on some of the same plant species as forage (Hardin et al. 1984) and cover. McKnight and Hepp (1995) found that grass carp in Alabama's Guntersville Reservoir foraged heavily on native muskgrass (*Chara* spp.) and southern naiad (*Najas guadalupensis*) in some areas, while the exotic species watermilfoil (*Myriophyllum spicatum*) was not strongly affected. They concluded that using grass carp to control aquatic vegetation may affect the overall habitat quality of reservoirs for waterfowl and waterbirds if the invading vegetation is a species that is not preferred by carp, such as watermilfoil. The decline in habitat quality results from the preference carp may show for the native species of vegetation in these systems, thereby also damaging habitat components essential to waterfowl.

Exotic species may also threaten biodiversity by altering the composition, structure, and ecological processes that characterize ecosystems. In Arizona, where exotic African lovegrasses (*Eragrostis* spp.) has been planted to revegetate degraded rangelands and provide cattle forage, Bock et al. (1986) found that regions dominated by these exotic grasses have a lower abundance and variety of native plants and wildlife than the native grassland community. Eight species of grasshoppers in Bock et al.'s study were significantly less abundant on exotic grasslands than on native grasslands, and the overall abundance of grasshoppers was 44 percent lower on exotic grasslands. Bird species abundance was also lower on exotic grasslands during the breeding season and winter. The theory proposed for the differences between the two grassland communities was that wildlife have evolved specific relationships within the native grassland community, and that the exotic vegetation does not provide the habitat attributes that will support those evolutionary relationships.

The invasion of flammable exotic plant species into various types of ecosystems is a serious problem recognized by ecologists worldwide (D'Antonio and Vitousek 1992). Two such species are cheatgrass, which threatens sagebrush (*Artemisia tridentata*) ecosystems in the western United States, and melaleuca, which is invading extensive areas of the Everglades. Melaleuca is an Australian tree species that tends to burn hot because of its high oil content. When melaleuca burns, the hot fires often kill seeds of existing native plant species while its own seeds survive. Following fire, dense stands of melaleuca typically develop, choking out native plant species. As melaleuca becomes established, it also can contribute to the drying of marshlands due to its ability to transpire relatively large volumes of water. If stands of melaleuca cannot be controlled, the biodiversity associated with regions of the Everglades will continue to be altered dramatically as native species are lost and disturbance regimes shifted (Jenkins 1996).

The fourth type of negative effect associated with exotic species is the transmission of diseases and parasites to native organisms. This is perhaps one of the most difficult threats to predict because exotic diseases and parasites can have countless unknown interactions within an ecosystem. Native species through evolution have developed resistance to a cohort of diseases and parasites occurring under a broad range of environmental conditions. However, if exotic diseases or parasites find suitable hosts, they may become established and have deleterious impacts on native biodiversity. Historically, such negative effects of diseases and parasites on biodiversity have been observed with the white pine blister rust (*Cronartium ribicola*), Port Orford cedar root rot disease (*Phytophthora lateralis*), Dutch elm disease (*Optostoma ulmi*), chestnut blight, and Marek disease.

The American chestnut (*Castanea dentata*) was a major component of eastern deciduous forests in North America until the early part of the twentieth century. During this period, chestnut blight (*Cryphonectria parasitica*), a fungal disease, was introduced on nursery stock from Asia, where it is endemic. The introduction of this disease virtually drove the American chestnut to extinction. Along with the near-extinction of the chestnut came the extinction of seven species of *Lepidoptera* that apparently fed exclusively on chestnuts (Opler 1978).

An outbreak of the avian disease known as Marek disease occurred in 1995 on the Galapagos Islands. This disease is believed to have been carried by chickens brought over from the Ecuador mainland. The relative isolation of the Galapagos Islands and the high proportion of endemic plants and animals that occur there make the ecosystem especially vulnerable to diseases introduced by exotic species, with potentially catastrophic consequences (Mauchamp 1997). Another example involves the common perch (*Perca flaviatilis*), which is native to Europe but was introduced to Australia in 1861 as a sportfish. The common perch is affected in the wild by epizootic hematopoietic necrosis, which has spread to cultured stocks of rainbow trout (*Oncorhynchus mykiss*) and to several native Australian fish species (Lever 1996).

Exotic species can be particularly threatening to biodiversity when they reduce the biological distinctiveness of an ecosystem with a unique species composition. Biodiversity does not refer solely to diversity of species but also to the diversity of communities and ecosystems, and it is not necessarily enhanced by increasing the number of species that occur in an area. Maintaining the assemblages of species that distinguish one native ecosystem from another is important for maintaining biodiversity beyond the species level, so the introduction of exotic species has the potential to blur or eliminate the distinctions among ecosystems.

One such example is the Great Lakes Basin in the United States. Coon (in press) reported that the Laurentian fish fauna has changed because of introductions of species to the basin and other human activities. The entire Laurentian fauna includes 190 species. Over the past 150 years introductions of at least 24 species that were not native to the Great Lakes Basin have been documented. These 24 exotic species are in the Cobitidae, Poeciliidae, Gobiidae, and Pleuronectidae families (Coon, in press). For instance, loach (*Misgurnus anguillicaudatus*), a fish produced for the aquarium trade, was accidently released from a captive population (Schultz 1960); bait fish species (e.g., ghost shiner [*Notropis buchanani*]) were established (Holm and Coker 1981); sea lamprey (*Petromyzon marinus*) were allowed passage through shipping canals (Lawrie 1970); flatfish (*Platichthys flesus*) were released along with bilge water from ships (Crossman 1991); and king salmon (*Oncorhynchus tshawytscha*) are intentionally released by humans (Hubbs and Lagler 1947; Coon, in press).

The combined releases of exotic species have had major effects on the fish community of the Great Lakes. The original food webs have been altered and now include many exotic species; moreover, some native species have been lost. Coon (in press) reported that eight native species have been completely extirpated from the basin.

Positive Effects of Exotic Species in Natural Ecosystems

Although people often may not realize it, not all introductions of exotic species succeed, and not all exotic species have negative effects on communities or other species. For example, planting exotic grasses when native species are not available may directly benefit biological diversity by providing wildlife habitat or by stabilizing a degraded landscape under restoration.

The Conservation Reserve Program (CRP), created by the 1985 Food Security Act, provides an illustration of how exotic plants have assisted in conserving biodiversity in grassland ecosystems. This program provides economic incentives to farmers to remove highly erodible cropland from production for ten years. The majority (87 percent) of these lands have been planted in grasses, an estimated 67 percent of which are exotic species (Osborn and Heimlich 1994). Best et al. (1997) have documented that bird abundance and nesting activity are greater in CRP fields than croplands across six midwestern states (Indiana, Iowa, Kansas, Michigan, Missouri, and Nebraska). Three of the species detected only on CRP fields (Henslow's sparrow [*Ammodramus henslowi*], bobolink [*Dolichonyx oryzivorus*], and sedge wren [*Cistothorus platensis*]) were identified by Herkert et al. (1996) as grassland birds of conservation concern. In addition, dickcissels (*Spiza americana*) and grasshopper sparrows (*Ammodramus savannarum*) commonly nested and were observed primarily on CRP lands (versus rowcrop fields) in all states except Michigan (Best et al. 1997). Five of the states participating in Best et al.'s study had fields planted in permanent exotic grasses and legumes (CP1 treatment). Landowner use of exotic grasses versus native species may have been an artifact of seed availability and cost, recommendations made by natural resource managers, and lack of awareness of the benefits of using native species (H. Campa, personal observation). Best et al. (1997) concluded that farm set-aside programs, such as CRP, that provide permanent grassland cover can be beneficial for conserving grassland birds.

Delisle and Savidge (1997) compared the vegetation and avian use of CRP fields in Nebraska seeded with exotic grasses and legumes (CP1) to those seeded with native grasses (CP2 treatment). Delisle and Savidge (1997) found that while some bird species showed a preference for either CP1 or CP2 plantings, total bird abundance did not differ between the types of plantings. Allen (1993) recommended that if a limited number of agricultural fields were to remain in the CRP, native grass plantings should be preferred over exotic plantings. An objective of planting and maintaining native grasslands, instead of grasslands dominated by exotic species, is to minimize some of the potential negative impacts that exotic species can have on the various characteristics of native ecosystems. In addition, Delisle and Savidge (1997) recommended that native grass plantings could provide habitat for many native grassland bird species if fields are managed by emulating natural disturbance regimes so that they provide the structural requirements native grassland birds require.

Recommendations for Managing Exotic Species for the Conservation of Biodiversity

Managing the spread of exotic species is a challenge for natural resource managers. Managers have little guidance about which exotic species will spread and what their effects will be on biodiversity. There are, however, some general characteristics of exotic species, and communities suseptible to invasion, summarized by Meffe and Carroll (1997), that managers should be aware of when making management decisions.

Generally, natural resource managers should look at the ecology of ecosystems and individual species that may be in question. For example, each species has a different physiology, behaviors, genetic composition, and ecology. Exotic species that have characteristics similar to native species may pose potential risks to native species because they are likely to compete. Also, managers should identify what types of ecosystems are more susceptible to the invasion of exotic species. Relatively fragile ecosystems that have not evolved with a species similar to an invading exotic may undergo significant ecological changes if an exotic becomes established, such as an exotic herbivore becoming established on islands.

Some species may have specific characteristics that tend to make them better invaders. Reichard and Hamilton (1997) investigated which plant traits were positively correlated with the invasiveness of species and whether they were related to the underlying causative characteristics of invasion. One of the most reliable variables in Reichard and Hamilton's models for predicting invasiveness of species was whether a species had invaded other areas around the world.

To minimize the potential deleterious impacts of exotic species on biological diversity, existing policies need to be revisited, as mentioned by Ruesink et al. (1995), and new ones developed. For example, regulations in the National Forest Management Act mention the maintenance of "native and desired nonnative species." Are these desired nonnative species having negative effects on biological diversity, or could they in the future? In the absence of a definitive answer, new policies need to specify that exotic species should not be used in habitat management activities where native species could be used, and that exotic species should not be intentionally released without careful consideration and study.

Some management guidelines and approaches (Ruesink et al. 1995) for managing exotic species to minimize their impacts on biodiversity may include the following:

- Development of more effective quarantines and surveillance at potential points of entry into ecosystems. These should be coupled with interstate and international agreements among natural resource agencies to minimize the establishment of exotic species and the deleterious effects exotics can have on native species and ecosystems.

- Testing an exotic species proposed for introduction for its potential invasiveness and direct threat to native species and ecological processes. These should be both short- and long-term concerns. While the introduction of a species may not show any negative short-term effects on native species or specific ecosystems, the long-term effects could be substantial. This management guideline would require that natural resource managers do a substantial amount of homework before intentionally introducing a species. Identifying whether an exotic species will be noninvasive and nonthreatening to other species and ecosystems is a daunting task. Determining if an exotic will be invasive in the short or long term will require a more sophisticated knowledge of the structure and function of ecosystems and how exotics as well as native species respond to disturbances and successional change. In addition, a regulatory body would have to be identified to evaluate the quality of the scientific investigations conducted on the proposed exotic species.

- The standards of states in the U.S. vary regarding what species and groups are regulated as exotic species and how carefully these standards are regulated. The Lacey Act (1900) leaves decisions on almost all intentional introductions of fish and wildlife species to the states, and only the few organisms on the list of injurious wildlife species are prohibited by the

Lacey Act (Office of Technology Assessment 1993). Consequently, a set of national minimum standards should probably be developed and enforced by state agencies.

- Management of ecosystems for conditions that do not allow exotic species to perpetuate themselves or invade areas. This may include avoiding management regimes that alter the composition and structure of ecosystems beyond their historical range of ecological conditions. Aplet and Keeton (Chapter 5) note that if ecosystems can be maintained within their historical range of variability, then perhaps there is a better probability of maintaining the composition, structure, and functions of natural ecosystems, thereby minimizing the invasions of exotics.

- Management practices that emulate natural disturbances while taking into the account their spatial scales as well as the temporal patterns of when they should occur. Initiating management practices within the time frames and at spatial scales that ecosystems evolved could aid in maintaining adequate habitat conditions for native species while creating unfavorable conditions for exotics.

- More drastic measures, such as culling or application of herbicide, if relatively conservative management activities (those that emulate natural disturbances, for example) do not control the numbers or spread of exotic species. These measures may be required if the natural functions of an ecosystem are in jeopardy and cannot be restored by other means.

- In some cases where exotic species have had undesirable effects, habitat management to mitigate their effects rather than to eliminate the species may be more appropriate and feasible.

Summary and Conclusions

In all their efforts, no matter what approach they take, natural resource managers must be aware of the many threats exotics can have on the conservation of biological diversity. Because of these threats, efforts must be made to use management practices that foster the maintenance of native species and functioning native ecosystems and that do not allow for the intrusion or propagation of exotics. Although we have presented several examples of how exotic species invade areas and the impacts they can have on the biodiversity associated with various ecosystems, there are no clear-cut ways by which to predict those impacts.

Careful evaluation is also necessary when intentionally introducing a species, and responsive management may be needed when an exotic invades or is intentionally introduced to an area. While there are many situations where the elimination of exotics would enhance or conserve biodiversity, elimination of all exotic species is neither a desirable nor a practical management goal.

Summary and Recommendations

This final section summarizes the information presented in earlier chapters. Our goal here has been to provide natural resource managers with straightforward and progressive advice in an easy-to-understand and -use format.

Chapter 14 reviews approaches and sets standardized criteria for managers to use in selecting the best one for their own particular management situation. In doing so, the author highlights the main principles that will enable natural resource managers to best meet biological diversity objectives. For example, to effectively maintain biological diversity at the landscape scale, managers must use ecological boundaries (across landownership boundaries) to explore potential patterns of diversity. When exploring and using ecological boundaries in decision making, managers must also use information on ecological processes and structure that help shape the potential diversity of landscapes. A species-by-species approach to plan for and manage biological diversity by itself is, in most cases, impractical. For this reason, natural resource managers may choose to also apply planning processes that integrate a continuum of coarse- and fine-filter levels of analysis, although by themselves both approaches have their limitations. Hierarchical approaches offer perhaps the best opportunity to conserve biological diversity across all levels of biological organization. In addition, maintenance of key landscape features that historically provided a diversity of habitat conditions for various organisms is suggested in certain situations. In the development of all strategies, the author envisions adaptive resource management as a necessary component in conserving biological diversity.

Chapter 15 describes a process that can assist managers in implementing strategies to conserve biodiversity. Kernohan and Haufler identify six steps, from identifying objectives to monitoring in an adaptive framework, that can serve as a key to success in conserving biodiversity. The authors project the image that managers need to pay attention to the sequence of implementation procedures, as well as learn from successful examples. Through these efforts, appropriate concerns can be allayed and strong partnerships can develop. Allowing a range of stakeholders to be involved in the decision making similarly helps to cement positive relationships. Implementation of an effective process for the conservation of biological diversity is about understanding the ecology of the system within today's political environment.

Future innovations and research needs that will allow natural resources managers to more effectively manage landscapes to conserve biological diversity are discussed in Chapter 16. Campa et al. develop their ideas in relation to suggestions and deficiencies identified in previous chapters. This chapter provides a realistic, prioritized amalgamation of research needs and potential management changes rather than a "wish list" of all possible research undertakings. The basis for the recommendations is the need for sound scientific analyses of current and future management approaches to conserve biological diversity.

Natural resource managers should not compromise their attempts to describe variations in diversity across landscapes by a lack of adequate replication or controls. A better understanding, through research, of the historic range of variability in ecosystem characteristics is promoted. Indeed, this understanding is essential if natural resource managers want to maintain a diversity of habitat conditions for a variety of organisms. The authors therefore encourage the use of computerized management tools and wildlife models to evaluate current and potential impacts of management practices on landscape composition and structure, availability of resources, wildlife populations, and habitat quality. Sensitive and flexible analytical techniques for evaluating the effects of management practices to conserve biological diversity among multiple scales (space, time, levels of biological organization) are also suggested. Finally, the authors advocate the accelerated integration of GIS coverages and forest and wildlife models to enhance managerial decision making.

Chapter 14

Contrasting Approaches for the Conservation of Biological Diversity

Jonathan B. Haufler

The various chapters in this book as well as additional literature describe strategies for the conservation of biodiversity. The assumptions and rationale underlying these approaches differ to varying degrees. This chapter identifies some of their advantages and disadvantages and issues to consider in their practical application.

Bioreserves

Numerous publications have proposed the use of bioreserves to maintain or enhance biodiversity (Harris 1984; McMahon 1993; Alverson et al. 1994; Noss 1994; Noss and Cooperrider 1994; Blockstein 1995; DellaSala et al. 1995, 1996; Cooperrider et al. Chapter 3). Under this strategy, areas of appropriate size and location are set aside as biological reserves, buffers, or zones of connectivity in order to protect species and ecological communities from negative human activities. Bioreserves can serve additional functions independent of their use as a strategy for conservation of biodiversity, such as reference conditions or educational purposes.

An advantage of the bioreserve strategy is that it is conceptually easy to describe—that is, the reserves and associated corridors can be delineated and displayed on a map. In reality, this delineation can require considerable thought and planning. Another advantage is that the idea of "protecting" areas is attractive to a segment of the public, as evidenced by the Wildlands Project (Davis 1994).

Nevertheless, the use of the bioreserve strategy for the purpose of "protecting" delineated areas from human intrusion and maintaining these areas in their current condition has its critics. Botkin (1990), Kimmins (1992), and others have argued that the bioreserve strategy can have only limited effects on biodiversity because of the dynamic nature of ecosystems and the unfeasibility of maintaining any system in a static state. In their view, as described by Salwasser et al. (1993:72), "[I]t is not possible to draw a line around an ecosystem and mandate that it stay the same or stay in place for all time. Managing ecosystems means working with the processes that cause them to vary and to change."

For this reason, Cooperrider et al. (Chapter 3) and Noss and Cooperrider (1994) argued that disturbance is an essential component in the maintenance of biodiversity. Noss and Cooperrider (1994) supported the establishment of a network of large bioreserves in which natural disturbance regimes are allowed to function, although they recognize the difficulties of allowing natural disturbances to occur in the presence of human development. They therefore propose that "most regions will require protection of some 25 to 75 percent of their total land area in core reserves and inner buffer zones, assuming that this acreage is distributed optimally with regard to representation of biodiversity and viability of species and well connected within the region and to other reserve networks in neighboring regions" (Noss and Cooperrider 1994:168).

The bioreserve strategy usually strives to set aside representative areas of all ecosystems in a reserve (Cooperrider et al. Chapter 3; Noss 1990; McMahon 1993). For this effort to be successful in conserving biodiversity, it must have a properly classified landscape for the "ecosystems" it strives to represent, and it must identify the correct numbers, sizes, and distribution of each type of ecosystem. In Chapter 3, Cooperrider et al. discuss the representation of all ecosystem types but do not identify a classification system that will accomplish it. In addition, the need for a process for identifying what constitutes an adequate amount and distribution of reserves to maintain biodiversity complicates the application of the bioreserve strategy without a process in place, there is no way of evaluating if biodiversity will be conserved.

Noss and Cooperrider (1994) primarily focused on mature or old-growth forested areas in discussing bioreserves. To counter the problem of fragmentation—that is, islands of protected land—they proposed that reserves be connected by corridors, such as forested riparian zones or wide, protected linkages containing "an adequate amount of interior habitat" (Noss and Cooperrider 1994:152).

The need for varied successional stages to preserve biodiversity is clear, but we do not yet know how early successional stages or grassland or shrub ecosystems should be intermixed and linked within forest ecosystems. Alverson et al. (1994), for instance, presented arguments for the use of late-successional corridors to maintain genetic interchange of bioreserve-dependent metapopulations, but the problem is, how will metapopulations dependent on earlier successional stages, such as those of sharptailed grouse (*Tympanuchus phasianellus*) or Karner blue butterflies (*Lycaeides melissa samuelis*), be linked to maintain their genetic interchange? Moreover, Camp et al. (1997) and Oliver et al. (1997) presented views on how landscape dynamics shift on a temporal scale, thereby creating, at different times, refugia for either early- or late-successional species. These temporal changes reveal the difficulty in the use of the strategy for maintaining metapopulations.

Kaufmann et al. (1994) discussed how ecosystem management attempts to meet ecological, social, and economic objectives. The bioreserve approach assigns these objectives to different areas within a landscape. While not a direct influence on the bioreserve approach for conserving biodiversity, this zoning of landscapes does have significant implications for the acceptance and implementation of the approach. For example, Alverson et al. (1994), Noss and Cooperrider (1994), and McMahon (1993) implied that core areas in the bioreserve strategy should be excluded from human influences and that commodity production is generally incompatible with their protection. By extension, we can say that the bioreserve strategy assumes that the production of commodities occurs predominantly in lands outside of core reserves; in other words, the production of commodities and maintenance of biodiversity in core reserve areas are mutually exclusive objectives. While few would argue that some types of human activities and some components of biodiversity cannot be maintained in the same place at the same time, the bioreserve strategy pursues the two objectives in different designated areas.

Scott et al. (Chapter 4) discuss the use of four categories of land ownership for the purposes of biodiversity conservation. Status 1 and Status 2 are areas such as national parks and wilderness areas. They advocate using a map of species distribution overlaid with a categorized land ownership map to identify needed areas for biodiversity prioritization and protection.

Emphasis-Area Strategy

Everett and Lehmkuhl (1996; Chapter 6) proposed a land management strategy termed an "emphasis-area approach." Under this approach, key areas of the landscape that support sensitive species or unique habitats are identified and designated as emphasis areas. These areas are then managed to maintain desired ecological values, flexible boundaries, and the dynamic nature that is characteristic of sustainable ecosystems. This strategy reduces what Everett and Lehmkuhl (1996; Chapter 6) termed "administrative fragmentation" and works to avoid setting aside areas that may not necessarily enhance the dynamics of a sustainable ecosystem. It also allows management activities, including commodity production, as long as the activities are designed to maintain or enhance the desired ecological conditions within the emphasis areas.

Emphasis areas provide considerably more flexibility in their application and management than bioreserves. By focusing on the ecological features of value in each emphasis area and the requirements for maintaining them, the emphasis-area strategy avoids the problems associated with a "protection" strategy that may not maintain the desired ecological feature over time. However, emphasis areas still share some of the same problems as bioreserves in conserving biodiversity because they focus on specific parts of a landscape. As with bioreserves, the approach does not in itself offer a mechanism for assessing whether biodiversity at a landscape scale will be maintained or enhanced in the emphasis areas. How many and what size of emphasis areas are needed could be difficult or impossible to quantify without an appropriate assessment of biodiversity. Focusing attention on only a limited part of the landscape may also ignore the contributions to regional biodiversity being provided by lands outside of the emphasis areas.

Coarse-Filter Strategies

Coarse-filter strategies have the goal of providing ecosystem integrity or biodiversity by maintaining the necessary mix of ecosystems at an appropriate landscape scale (Kaufmann et al. 1994). These strategies assume that biodiversity and ecosystem integrity can be maintained if the correct mix of ecosystems or ecological communities is provided.

All coarse-filter strategies are based on a classification system of a landscape. For this classification system to function correctly as a coarse filter, it must partition the landscape into meaningful ecological communities. The classification system should include sufficient numbers of ecological communities so that if each community of the classification is provided, biodiversity will be maintained. In contrast, if the classification is too detailed, then unneeded complexity is incorporated into the management planning process. Coarse-filter strategies generally include contributions toward biodiversity from the entire planning landscape.

Strategies Based on Habitat Diversity

Oliver (1992) discussed a strategy termed "landscape ecosystem management." In this strategy, described in Chapter 3, active management is used to maintain a mix of four stand structures across a landscape. The approach is a coarse-filter strategy based on habitat diversity represented by the four structural stages.

Landscape ecosystem management (Oliver 1992) is comparatively easy to apply because it requires that managers keep track of and plan for a mix of stand structures across an area. The method emphasizes the dynamic nature of ecosystems over time. Landscape ecosystem management also provides for management to meet a variety of human needs, including commodity production. It thus meets the ecosystem management goal of balancing ecological objectives with human values and needs.

But the use of landscape ecosystem management as an approach for biodiversity conservation raises several issues. First, as noted earlier, the success of any habitat diversity approach for conserving biological diversity lies in how the landscape is classified and whether or not the classification system properly partitions the landscape into an appropriate coarse filter. Management for four stand structures, without regard to a landscape's additional ecological complexity, may not adequately address the range of ecological communities needed for conservation of biological diversity. For example, under this approach, an old-growth northern hardwood stand would be considered the same as an old-growth northern white cedar (*Thuja occi-*

dentalis) stand. Similarly, an old-growth Sitka spruce (*Picea sitchensis*) forest along the Pacific Northwest coast would be classified the same as a higher-elevation, old-growth mountain hemlock (*Tsuga mertensiana*) forest. The ecological complexity among these different forest communities may be critical to planning for either specific species or for biodiversity. Thus, the four structural stages proposed by Oliver (1992) may not classify a landscape in sufficient detail to account for the full range of biological diversity.

Ecological classifications such as habitat typing (Steele et al. 1981; Pfister 1993) and the U.S. Forest Service's ecological classification system (Jones and Lloyd 1993; Ecomap 1993; Roloff 1994) are classification systems for delineating the ecological complexity of a landscape. The addition of an ecological classification to Oliver's (1992) stand structures addresses the concern in landscape ecosystem management for the need to consider the ecological complexity of a landscape.

A second issue regarding the use of four stand structures is whether these adequately describe the complex of successional stages that support biodiversity. It could be assumed that if the four different structures were maintained, additional ecological communities representing additional conditions among the four structures would also be maintained. However, if finer classifications of successional change were identified as important, the results would need to be closely monitored.

A concern with all habitat diversity approaches is that the classification of a landscape based on a generalized sequence of habitat change may not emphasize ecological communities that occurred under historical disturbance regimes. In other words, a classification for habitat diversity may require, for example, seven different structural stages (O'Hara et al. 1996) to be maintained. However, for many habitat types (Steele et al. 1981) in the landscape, only four or five of these structural stages may have occurred. Under landscape ecosystem management, the additional two or three stages would need to be maintained, even though historically no species depended on them, potentially adding unneeded restrictions on land management planning.

The biggest criticism of any habitat diversity approach used on its own is that the needs of some species may not be met through the maintenance of a mix of successional stages or stand structures (Kaufmann et al. 1994). Certain species with specific stenotrophic habitat requirements could "slip through the cracks" in a habitat diversity approach, and the problem would not be detected until the species were in peril. In particular, species that may require specific spatial configurations of habitat attributes provided by several biological communities may not be provided for in a habitat diversity approach. Also, assumptions about amounts of specific stand structures or successional stages needed to maintain biodiversity must be validated. If such approaches are combined with an adequate method of selected species assessments or evaluations, these concerns may be addressed.

Strategies Based on Historical Range of Variability

Another coarse-filter strategy for biodiversity conservation discussed in Chapters 2 and 5 is management for historical range of variability. The assumption of this approach, that species native to an area evolved with and adapted to conditions created and maintained by historical disturbance regimes, is one of its strengths. It is difficult to argue against historical landscape functions, structures, and compositions as a solid basis for biodiversity conservation. This is the greatest advantage of this approach. Identification of historical range of variability provides an understanding of the landscape that can serve as a reference point for assessing human-induced change in landscapes.

An argument against the use of historical range of variability is that science has not been able to adequately define many specific historical ranges of variability. Management targeted to produce desired stand conditions may not achieve landscapes within the historical range because of a lack of suitable definition. Considerable discussion has occurred over what is an appropriate target for historical range of variability—whether it should be 100 years ago, 400 years ago, 2,000 years ago, or even earlier.

The techniques available for determining historical range of variability are inadequate to make accurate determinations in many ecosystems. Systems in which vegetation had rapid decomposition rates or that were typically subjected to stand-replacing fires may not leave reliable evidence of disturbances, as might be provided by fire scars in ecosystems subjected to understory fire regimes. Thus, while the scientific basis for historical disturbance regimes is perhaps the strongest of any approach, the accurate determination of historical range of variability is still problematic.

The dynamic nature of ecosystems is also a factor to be considered in applying historical range of variability. Species have shifted distributions within time frames considered as targets for historical ranges of variability, and these shifts have caused changes in ecosystem composition, structure, and processes. Davis et al. (1994) discussed the effects of the invasion of hemlock (*Tsuga canadensis*) into northern Michigan 3,200 years ago and the change in the forests that resulted from this invasion. Similarly, changing climatic patterns and shifts in species distributions that may occur as a result raise concerns as to the long-term effectiveness of managing an area within its historical conditions. Even leaving stands alone or using natural tools such as fire may not produce the desired results because of past human influences such as unnatural fuel accumulations or changes in species compositions. For example, the American chestnut (*Castanea dentata*) was an important species in many eastern U.S. forests, but at present cannot be returned to its previous role in these ecosystems. The influence of indigenous peoples in structuring the historical landscape around themselves is also receiving increased attention as we work to understand how humans have influenced their environments. On the basis of these concerns, management for historical ranges of variability may not provide for the needs of certain species and, thus, may not maintain biodiversity.

An additional issue is that management within historical ranges of variability greatly restricts the range of management alternatives and may not adequately meet human values and needs. While managing landscapes outside of historical ranges of variability may increase potential risks to ecosystem integrity (Morgan et al. 1994), it is thought to be feasible to maintain biodiversity within a landscape while outside of these ranges. In fact, in most landscapes of the country we are already considerably outside of historical ranges of variability for many ecosystem variables and cannot return to historical ranges without substantial alterations to the human population present. Thus, while potentially a viable alternative in some areas from an ecological perspective, a strict goal of maintaining a landscape within historical ranges of variability is generally unrealistic or impossible to achieve from the perspective of human needs or economics.

Aplet and Keeton (Chapter 5) discussed the use of historical range of variability in identifying ecological communities for representation in bioreserves. They further described how an understanding of historical range of variability can be used to assist in determining sizes of representative stands and temporal connectivity of stands. But again, as with the strategies discussed earlier, there are no mechanisms for determining effectiveness—in this case, what constitutes adequate representation of ecological communities. While the assumption that maintaining a landscape within historical ranges of variability will maintain biodiversity has a good scientific basis, maintaining landscapes under this approach without an evaluative mechanism may not meet biodiversity conservation objectives.

Fine-Filter Strategies

Fine-filter strategies focus on providing habitat conditions for the specific needs of individual species, guilds, or other groupings of species. Wall (Chapter 8) discussed a landscape planning strategy designed to provide habitat conditions for guilds of species. Wall did not assert that the fine-filter approach he described would maintain all biodiversity in his planning landscape; rather, he believed that it would be valuable for predicting species assemblages spatially and temporally.

Fine-filter approaches offer several advantages. They do address the needs of the Endangered Species Act (ESA) and the species viability components of the National Forest Management Act (NFMA). Agencies including the U.S. Fish and Wildlife Service and the U.S. Forest Service are required by legislation to include some focus on species in their management responsibilities. Thus, even if they utilize a coarse-filter approach for addressing ecological objectives, they must also assess the needs of selected species.

Fine-filter approaches also address the concerns of the public for protection of selected species. For example, large segments of the public may have specific concerns about wolves (*Canis lupus*) that will not be specifically addressed by a coarse-filter strategy. Fine-filter approaches have the advantage of considering the contribution of the entire planning landscape to biodiversity, not just selected areas as in the bioreserve approach. Finally, fine-filter approaches can be designed to balance the needs of species with human demands.

Again, lack of information is a concern. One problem with fine-filter strategies is they assume that a sufficient knowledge base exists to adequately describe the habitat requirements of all species potentially occurring across the landscape. While habitat requirements of some species are known, for other species, specific habitat requirements remain unknown or are known only in the most general terms. For fine-filter strategies to work effectively, a large database on habitat characteristics of the landscape may be needed. This database will require a mapped landscape classification system of some type to account for the quantity, quality, or distribution of habitat for the different species.

While maintaining viable populations of species is an essential requirement of all strategies for ecosystem management, fine-filter strategies focus on meeting specific needs of species or species groups and determine whether these are provided for across the planning landscape. Selecting indicator species or using species guilds is an attempt to simplify the process of meeting the needs of the numerous species that occur in a landscape. The difficulty in using guilds or indicator species is that the specific needs of many species are not addressed, and concerns can be raised as to the adequacy of the plan to maintain all species. Morrison et al. (1992:246) have noted, "Whereas grouping species by guilds might be useful for depicting groups of species with similar functions, the guild approach is not useful for predicting specific responses of each species to environmental conditions and changes. To adequately perform the latter task, a researcher could model each species within the guild individually and then combine results." Hunter (1990) stated similar concerns that guilds do not address the needs of individual species. Thus, species or finer stratification of guilds or groupings can be added to the fine filter until concerns about each individual species are addressed, but then many species must be considered. For example, the SAT report (Thomas et al. 1993) identified 517 species dependent on late-successional stages alone. Trying to balance the conflicting demands of all these species at a landscape scale becomes a daunting task. Marcot et al. (1994) stated that a weak point of this strategy is in meeting the conflicting needs of disparate species. Risbrudt (1992) pointed out that a species-by-species approach is not feasible because of the large number and complexity of species involved.

A final criticism of the fine-filter approach is that it does not incorporate the consideration of ecological processes in its assessment and planning of landscapes. Such factors as disturbance regimes and ecosystem functions may be critical for the long-term conservation of biological diversity but may be completely overlooked in a fine-filter approach. Thus, the fine-filter approach has some major implementation problems.

Coarse-Filter Approach with a Species Assessment

The coarse-filter approach with a species assessment (Haufler et al. 1996; Chapter 7) uses a coarse filter based on historical ranges of variability; it then identifies what is termed "adequate ecological representation" to meet ecological objectives. In addition, the approach provides a mechanism to check the adequate ecological representation through assessment of the viability of selected species, to minimize the chance that species might slip through the cracks of the coarse-filter approach. Hunter (1990) and Kaufmann et al. (1994) recognized the advantages of this type of combined approach, though they did not identify a framework for its application.

This strategy has a number of benefits to recommend it. First, it has the operational advantage of using a coarse filter to characterize the landscape, avoiding the problem of tracking large numbers of species that is inherent in fine-filter approaches. In using a coarse filter, it allows for biodiversity contributions of the entire landscape to be included, rather than targeting only selected areas of the landscape. The coarse filter incorporates ecological processes and allows for ecosystem dynamics to be included in landscape planning. It utilizes historical ranges of variability as a basis for defining the coarse filter, but then it identifies adequate ecological representation rather than requiring the landscape to be within historical ranges of variability. It incorporates species assessments as a check to ensure that the coarse filter is working. The species assessments can also address the species requirements of ESA or NFMA as well as public concerns about specific species. Finally, it allows for the inclusion of economic and social concerns in the planning process, providing flexibility in landscape planning.

The coarse-filter approach with a species assessment does have disadvantages. It requires significant data to implement, in that the coarse-filter planning must incorporate sufficient fine-scale data to allow for an adequate assessment of any selected species. It requires historical ranges of variability to be understood and quantified, thus the concerns expressed earlier about defining historical ranges of variability. It requires the correct identification of adequate ecological representation of the coarse filter. It also relies upon the assumption that assessments of quantities and qualities of habitat can be linked to desired population responses by species targeted in the species assessment. To address this concern, actual population surveys could be incorporated into the species assessment, if needed, and certainly should be used to validate species models.

Broader-Scale and Restoration Efforts

Samson and Knopf (Chapter 10) addressed issues of biodiversity conservation at a larger scale than those discussed earlier. They identified the need to assess the distribution of endemics and the prioritization that must occur at regional scales if biodiversity conservation is to be successful. This prioritization needs to include the maintenance of native ecosystem processes that are essential for conserving biological diversity across landscapes.

Ncraasen and Nelson (Chapter 9) discussed some innovative methods of accomplishing ecosystem restoration in prairie landscapes that have undergone substantial conversion to agriculture. Their ability to accomplish this restoration in an area with large economic competition for land use and across a diversity of stakeholders is significant for effective application of any strategy. An important feature of the approach is the integration of scientific knowledge within a complex sociopolitical and values framework under the general goal of protecting ecosystem integrity over the long term. In addition, the application of adaptive resource management principles, especially with respect to monitoring and feedback, is central to the program's success.

Regional Approaches

Scott et al. (Chapter 4) discussed the use of Gap Analysis as the basis for regional planning efforts to conserve biodiversity. This method includes the construction of multiple regional data layers (land cover and the distribution of terrestrial vertebrates, for example). A goal of developing these data layers is to integrate them to aid in identifying key geographic areas of "high conservation value." This approach to conserving biodiversity has been used in selecting proposed sites for national parks, in reviewing alternative plans for enhancing diversity in wilderness areas, and as a methodology to describe ecoregions within states. In essence, the focus of this approach hinges on identifying "biologically significant areas" for protection.

Common Features of Approaches

Each of the strategies discussed in this book is distinct from the others, but some have features in common.

The bioreserve approach and the emphasis-use approach have many similar characteristics. If the more general definition of bioreserves as areas designated to conserve biodiversity is used, then these two approaches overlap. Under the more narrow definition of bioreserves that excludes, to the maximum extent possible, human activities, then these two diverge. As discussed by Noss and Cooperrider (1994) and others, recognition of the need for various types of human activities to obtain biodiversity objectives within many bioreserves may, in practical terms, move these two approaches closer together. However, there would still be a basic philosophical difference between the two approaches regarding how they would be incorporated into landscape planning efforts for biodiversity conservation.

In Chapter 3, Cooperrider et al. discussed the goal of obtaining representation of all ecological communities. As cited in Noss and Cooperrider (1994), this goal is "[t]o represent, in a system of protected areas, all native ecosystem types and seral stages across their natural range of variation." As such, it is very similar to the goal of a coarse-filter approach for historical ranges of variability (Aplet and Keeton, Chapter 5), and it is similar to that of the coarse-filter approach with a species assessment (Haufler et al. Chapter 7). However, the bioreserve approach differs from the coarse filter with a species assessment in the expectation that the represented communities are contained in bioreserves that minimize human activities. Haufler et al. do not place these constraints on the delineation of areas for adequate ecological representation.

The species assessment component of the approach of Haufler et al. (1996; Chapter 7) could be used as a check if it is included with other approaches. However, the other approaches would need to classify and map a planning landscape by means of methods that would supply habitat descriptions of the needed precision and accuracy to drive the species models (Roloff and Haufler 1997). Doing so would address concerns that the bioreserve, emphasis-use, and coarse-filter approaches based on habitat diversity or historical range of variability lack a mechanism to check their effectiveness in meeting biodiversity conservation objectives.

All of the above approaches will require new types of information and data applied at new, larger scales. Most of the approaches are working hypotheses and as such need monitoring and, potentially, adjustment. The adaptive management aspects are particularly emphasized by Wall (Chapter 8), Nudds (Chapter 11), and Kernohan and Haufler (Chapter 15).

While continued discussion and debate over the advantages, disadvantages, and appropriate applications of each of these proposed approaches will continue, it is important to note that all advocate the use of larger-scale landscape assessments and methodologies to meet the objectives of biodiversity conservation. A challenge is in achieving this objective in mixed ownership landscapes (Cooperrider et al., Chapter 3; Haufler et al., Chapter 7; Kernohan and Haufler, Chapter 15), and these authors stressed the need for collaborative efforts. The success of these collaborative efforts may largely depend on the acceptance of the proposed approach by a diverse group of participants. Cooperrider et al. (Chapter 3) discussed some of the complexity and challenges faced by a proposed bioreserve effort in the Klamath Ecoregion. Kernohan and Haufler (Chapter 15) discussed considerations in developing effective collaborative initiatives.

Implications

Each approach has its own particular philosophical basis and assumptions, so it must be kept in mind that the one selected will significantly influence the outcome. Thus, alternative management actions planned across a landscape can be strongly influenced and even constrained by the framework of the selected approach. Moreover, the design of alternative management plans once a conservation approach has been selected can be very different under the different approaches; consequently, consideration of the influence of the approach on the outcome of possible management alternatives is a critical step in a landscape planning process. Managers and planners must thus be aware of the constraints that selection of any one approach can place on the outcome of the planning process. Because each approach has associated advantages and disadvantages, it is appropriate to consider each one prior to implementing any biodiversity conservation effort.

Implementation of an Effective Process for the Conservation of Biological Diversity

Brian J. Kernohan and
Jonathan B. Haufler

All effective efforts at maintaining or enhancing biological diversity require a process to be followed in decision making. This chapter outlines a generalized process for conserving biological diversity and discusses considerations that are necessary to implement the process. The process we have identified includes

- identifying objectives and limiting factors
- identifying the extent of the planning landscape and plan duration
- selecting an appropriate classification system and mapping it at an appropriate resolution
- identifying biological diversity problems on the basis of limiting factors
- involving needed partnerships
- designing public relations and involvement steps
- monitoring in an adaptive management framework.

The process begins by describing specific objectives and limiting factors of the strategy under consideration. Depending on the strategy, appropriate spatial and temporal scales must be considered in delineating the planning landscape and the plan's duration. Within this landscape, an appropriate classification system must be identified or developed. All conservation strategies depend on the production of a land classification map in order to identify biological diversity problems on the basis of the chosen strategy's limiting factors. The map should be produced at an appropriate resolution. Following identification of problems, an appropriate partnership network that includes land management agencies, environmental organizations, private land owners, elected officials, and concerned citizens should be used. In addition, good communication and public awareness from the onset of conservation efforts are important to the successful achievement of the objectives. Finally, an appropriate adaptive management strategy should be employed to effectively monitor, evaluate, and adjust the process.

Implementing an effective process for the maintenance of biological diversity must consider and integrate ecological, political, economic, social, and legal issues. Ecological issues include spatial and temporal scales, ecological boundaries, land classification systems, and determination of appropriate distributions of biodiversity. Political concerns often overshadow ecological issues and include considerations such as political boundaries, agency-landowner responsibilities, funding potentials, and political agendas. However, such considerations need to be examined in the context of appropriate ecological objectives. Social and economic issues function similarly to political concerns with regard to how they should be considered in the context of ecological objectives. Such issues include natural-resource-based economies, employment opportunities, recreation, aesthetics, and cultural values. Finally, legal issues such as antitrust concerns must be considered when implementing programs for the maintenance of biological diversity.

As discussed in the chapters of this volume and other manuscripts, various strategies have been proposed for conserving biological diversity, including bioreserve strategies (Cooperider et al., Chapter 3; Scott et al., Chapter 4); coarse-filter approaches based on either historical range of variability (Aplet and Keeton, Chapter 5) or habitat diversity (Oliver 1992); fine-filter strategies (Wall, Chapter 9); and a coarse filter approach with a species assessment (Haufler et al., Chapter 7). The process we have identified applies to the implementation of any of these strategies, however, differences in application will exist and success will vary depending on overall conservation objective and degree of implementation. For purposes of this chapter, the process will be described in a general sense, and examples will be related to the strategies.

Describe Objectives and Identify Limiting Factors

One of the fundamental challenges facing the conservation of biological diversity is understanding how biodiversity should be measured and maintained across a planning landscape. Measures for conserving biodiversity vary depending on the conservation strategy chosen. An example of a measure for a fine-filter strategy may be an appropriate distribution of viable populations of all native flora and fauna in a planning landscape. Whereas coarse-filter approaches depend on providing the correct mix of ecological communities so the question to be answered is whether there is sufficient quality and quantity of needed ecological communities to provide for all native flora and fauna. Appropriate measures are different for each of the conservation strategies. Measurements of the success of coarse-filter approaches will consider whether they provide the correct mix of ecological communities.

Bioreserve strategies generally set planning objectives on the basis of protecting either "hot spots" of biodiversity on the landscape (Scott et al., Chapter 4) or representation of all successional stages of each plant community type (Cooperrider et al. Chapter 3), and excluding such reserves from human activities. Their objectives assume that either the "hot spots" or the represented successional stages maintain all native flora and fauna. Therefore, limiting factors to this approach can be identified as either a lack of protection for a sufficient number of hot spots or a lack of sufficient representation of successional stages in reserves.

Coarse-filter approaches generally describe ecologically meaningful communities for ecosystem integrity and species habitat. A coarse-filter strategy based on habitat diversity would have the objective of providing a mix of habitat conditions across the landscape that accommodates all native flora and fauna. How habitat diversity is defined is critical to the success of the approach. For example, Oliver (1992) described habitat diversity based on four structural stages. The assumption is that through provision of these four structural stages, biodiversity will be maintained. The limiting factor of this strategy would be not providing all four structural stages.

A second type of coarse-filter strategy uses historical range of variability to define habitat conditions. The assumption of this approach is that native species evolved with historical conditions; therefore, maintaining a landscape within historical ranges of variability will preserve biodiversity. This approach requires the measurement of landscape conditions in relation to defined historical ranges of variability. The limiting factor of this approach is evident when the quantity or distribution of ecological communities is outside historical ranges of variability.

A stated objective of fine-filter approaches is to provide habitat for all species, often through use of guilds or indicator species. Fine filters require habitat attributes to be described and mapped at appropriate resolutions to develop estimates of species occurrence or population viability. Roloff and Haufler (1997) described a means of assessing population viability of single species at a landscape level through measurement of habitat quality and quantity. Limiting factors of this type of approach can be described as inadequate habitat for any species in question.

Haufler et al. (Chapter 7) described an approach that combines a coarse filter defined as adequate amounts of inherent ecosystems with a species assessment as a check on the coarse filter. The approach employs an ecosystem diversity matrix as a tool to describe the landscape with regard to land potential and vegetation growth stages. The desired objective is maintenance of adequate ecological representation (that is, sufficient amount and distribution of inherent ecosystems to maintain viable populations of all native species [Haufler 1994]). Therefore, the limiting factor in this approach is a landscape lacking the threshold level for adequate ecological representation of one or more inherent ecosystems.

Thus, each approach has the same overall objective of conserving biological diversity; yet each has very different, specific objectives and different factors that will limit attainment of the objective. For effective implementation of each strategy, the first step is the clear statement of the objectives and limiting factors.

Identify Appropriate Spatial and Temporal Extent

One of the difficult challenges facing the maintenance of biological diversity is determining the appropriate extent of the landscape in terms of both spatial and temporal scales. How the complexity of spatial scales is organized is critical and is best achieved through application of hierarchical theory (Haufler et al., in press). Temporal scales must be considered in order to understand the duration of both natural and human-induced disturbances, successional trajectories, appropriate planning horizons, and the length of any historical perspectives (Haufler et al., in press).

Determining the appropriate spatial extent of the landscape is critical and must be ecologically based. Adequate maintenance of biodiversity generally requires management decisions over a large geographic area (Gregg 1994). Landscape objectives can be met only when they are framed within appropriate scales. Identification and delineation of an appropriate landscape should be based primarily on ecological objectives (Haufler et al. 1996; Chapter 7). Social and economic objectives operate at and should be defined at a variety of spatial scales and related to the confines of the ecologically defined landscape. Haufler et al. (1996) described four primary criteria that should be considered when delineating a planning landscape:

1. Similar biogeoclimatic conditions that influence site potentials.

2. Similar historical disturbance regimes that influence vegetation structures and species compositions.

3. Landscape of adequate size to provide sufficient ranges of habitat conditions to assure population maintenance of the majority of native species that historically occurred in the planning landscape.

4. Recognition of maximum size to avoid practical limitations of data management, implementation restrictions, and number of cooperating landowners necessary for successful plans. In addition, acceptable levels of variation in characteristics of ecological land units within the planning landscape must be understood to properly delineate a planning landscape (Haufler et al. Chapter 7). Although the recommendations proposed by Haufler et al. in Chapter 7 describe the delineation of a planning landscape for ecosystem management, the five criteria may also be applied to the definition of a landscape for the maintenance of biodiversity because maintenance of biodiversity is a specific objective of ecosystem management (Keystone 1996). The U.S. Forest Service's national hierarchy of ecological units (Ecomap 1993) and Maxwell et al. (1995) provided a hierarchical classification

that provides an ecological framework for boundary determination. Haufler et al. (1996), applying the above criteria and classification systems, advocated the use of the section or aggregates of subsections level (equivalent to Bailey's subregions [Bailey 1995, 1996]) to provide an appropriate landscape that would both meet ecological objectives and be operationally functional.

Most past management planning has utilized legal, political, and ownership boundaries as the basis for decisions, but these operate at multiple scales and can change within relatively short time intervals (one or two years). Social and economic objectives will usually reflect economic markets, political structures, and social influences (Haufler et al., in press). While these objectives must be considered, biodiversity maintenance must be based on ecological boundaries, since ecological boundaries defined by the integration of biogeoclimatic factors (such as climate, geology, and soils) remain static with respect to current planning time frames and operate at discrete levels.

These considerations for delineating a planning landscape apply to all of the five example strategies. However, in the actual implementation of most of these strategies, the extent of the planning landscape has not been determined using these considerations. For bioreserves, Cooperrider et al. (Chapter 3) discussed the Klamath Ecoregion as the planning landscape. For coarse-filter strategies, planning landscapes of various sizes have been suggested. Similarly, fine-filter strategies can operate in many-sized landscapes, but meeting species viability objectives may require large areas.

Maintaining biological diversity involves time frames that are often far beyond traditional planning horizons (Haufler et al., in press). Protecting genetic components of biological diversity may involve analysis and planning for multiple generations of a species to ensure that adequate heterozygosity of gene pools is maintained (Haufler et al., in press). Another example of the importance of temporal scales involves the exchange of genetic information and provisions for demographic and environmental stochasticity among metapopulations of a species. Temporal scale also must consider the time spans associated with the duration of natural disturbance regimes within the landscape. Planning time spans that do not sufficiently consider such events fall short of adequately maintaining biodiversity. Haufler et al. (in press) suggested that planning time spans recognize degrees of expectations at increasing time intervals. For example, relatively detailed plans are often expected for the immediate future, whereas less-detailed plans might be expected for a twenty-to-fifty-year horizon.

Identifying the extent of the planning time span is also often neglected in implementation of specific strategies. For bioreserves, the extent of the planning time span is assumed to be in perpetuity. Coarse-filter approaches require the consideration of successional change and should consider time spans that incorporate all successional stages. Fine-filter approaches need to consider changing habitat conditions but can be of varying time spans.

Identify and Map an Appropriate Classification System

All conservation strategies for biological diversity will require some type of map to implement required actions. All such maps are based on a classification system of land units. The appropriateness of a classification system will vary depending on the conservation strategy. Meanwhile, the resolution of mapped units in the classification system will influence the effectiveness of the effort relative to the strategy. Therefore, both the classification system used and the resolution of mapping are important considerations in implementing effective conservation strategies.

Bioreserves

Land classification systems required to implement bioreserve strategies can vary depending on the bioreserve approach being used. For example, Scott et al. (Chapter 4) determined "hot spots" based on habitat mapping of a large number of species. The classification system used needs to be able to describe habitat of these species and needs to be mapped at a sufficiently fine resolution to identify habitat patches for the selected species. Haufler et al. (in press) discussed resolution issues related to these mapping efforts. Cooperrider et al. (Chapter 3) required a classification system capable of mapping successional stages and potential natural vegetation as a coarse filter, although they did not specify the classification system or its mapping requirements.

Coarse Filter–Habitat Diversity

Oliver (1992) described a coarse-filter classification system that delineated stand structural stages (termed "landscape ecosystem management"). The resolution of the defined structural stages is capable of delineating landscape-level changes in vegetation structure only and does not consider site-level variability. Oliver (1992) proposed the maintenance of a mix of four stand structural stages across a landscape to maintain biological diversity, but this classification does not consider ecological site differences or the ecological complexity of the landscape. Moreover, Oliver (1992) did not discuss the mapping resolution needed to implement landscape ecosystem management, although a fairly fine scale resolution is assumed to be needed.

Coarse Filter–Historical Range of Variability

Coarse-filter classifications based on historical range of variability require classification of historical disturbance regimes and the communities they produced. Historical conditions have been mapped from General Land Office survey notes, fire scars, old photographs, and pollen analysis. Alternatively, historical disturbance regimes can be described to ecological sites, such as habitat types (Daubenmire 1968), and these types mapped across a landscape. These classifications must be of sufficient detail to distinguish among the different influences caused by different disturbance regimes. Mapping resolutions must be sufficient to properly place disturbance patterns across the planning landscape. In addition, existing conditions need to be mapped as a comparison to the desired historical ranges of variability.

Fine Filter

Fine-filter strategies require mapping of vegetation conditions in sufficient detail and resolution to describe habitat quantity or quality for selected species. The classification may be of vegetation structure or composition but must allow for adequate description of habitat attributes for species habitat assessments. For planning future conditions, the classification system should allow for the consideration of successional change. Classification systems that are based on dominant overstory vegetation alone may not adequately describe habitat features of the understory. Mapping resolutions must allow for delineation of homogeneous stands at scales utilized by the suite of species being evaluated.

Coarse Filter with a Species Assessment

The combination of existing vegetation structure and site potential produces an ecologically based classification system that will provide a description of uniform units based on predictable and repeatable site characteristics within the landscape (Russell and Jordan 1991). Haufler et al. (Chapter 7) defined ecological land units by combining vegetation growth stages (that is, units capable of differentiating the various biological communities represented within successional stages) with site potential. This classification allows for the incorporation of historical disturbance regimes, which in turn are used to identify inherent ecosystems. It also provides detail sufficient to map existing vegetation and to describe habitat features for selected species. This classification system must be mapped at a resolution that defines homogeneous habitat conditions for the selected species assessments.

Identify Biological Diversity Problems on the Basis of Limiting Factors

With clearly stated objectives, a chosen strategy, and an understanding of the limiting factors of the strategy, identification of biological diversity problems should be straightforward. This of course assumes that an appropriate classification system has been developed and mapped.

Problems identified under a bioreserve strategy may include a lack of protection for a particular community of high species diversity. One solution may be to generate a "set-aside" protection plan for the area to ensure that nothing disrupts the population. Identification of problems under either coarse-filter approach will likely result in review of whether or not an appropriate mix of habitat units, either habitat diversity units or units described by historical conditions, is provided for across the landscape. Problem identification when using a fine-filter strategy will likely consist of whether or not specific species habitat requirements are met. This will have to be done on a species-by-species basis; thus, problems may arise where habitat requirements for individual species contradict each other. Finally, if the coarse filter with a species assessment approach is used, problems are detected when individual ecological land units fall below adequate ecological representation, either in amount or distribution, across the landscape.

Determine How to Correct Problems through Use of a Partnership Approach

Collaboration among organizations and individuals is usually critical to the success of an ecosystem management initiative (Keystone 1996). The same principle holds true for initiatives directed to conservation of biological diversity. Using an effective collaborative process can minimize or even eliminate the inherent conflicts caused by changing legal, economic, or ecological landscapes that can derail or distort decision making (Keystone 1996).

Building successful partnerships is key to true collaboration. For partnerships to be accepted and successful, they must include the following: (1) voluntary involvement, (2) a broad range of participants, (3) involvement from the onset, (4) identification of mutual goals and objectives, (5) respect for individual landowner objectives, (6) decision making by consensus, and (7) trust.

Successful partnerships begin with voluntary participation by a broad range of participants (Haufler 1995; Keystone 1996). Participants should include county, state, and federal governments, industrial and nonindustrial private landowners, environmental organizations, tribal governments, elected officials, and concerned citizens. Once voluntary participants have been assembled, it is critical to define the participants' roles and authorities to enhance their involvement and to avoid misunderstandings (Keystone 1996). In addition to active participants, technical expertise and facilitation may be needed to assist in accomplishing the objectives of the partnership.

Ownership in the decision-making process from the beginning is critical and can be achieved only through involvement by all interested participants from the outset (Haufler 1995). Continuity of participation is critical; however, the time and resource limitations of participants need to be recognized (Keystone 1996).

For true collaboration to exist, potential partners cannot come to the table with preconceived agendas or methodologies (Haufler 1995). But even if true collaboration may never exist, it is important that stakeholders are flexible about their agendas and how they intend to implement them. Before any agenda is set, the partnership needs to decide upon mutual goals and objectives, or vision, and a process for achieving it.

Ground rules must be established before any implementation of strategies. One of the most important ground rules is to respect the objectives of all participants, including individual landowners (Haufler 1995; Keystone 1996). All participants come to the partnership with their own objectives; however, their objectives are not to be confused with the mutually agreed upon objectives and goals of the partnership. A participant must strive to meet the objectives of the group while ensuring that his or her own objectives are not compromised or lost in the process.

The goals of the partnership will be best met by means of a consensus process. Consensus means considering all points of view and working toward a solution. Care should be given not to confuse consensus with compromising individual landowner objectives. When differences exist, it is helpful for participants to agree to disagree and move on. The process of developing consensus will facilitate implementation of most initiatives and encourages participants to assist resource managers in generating support for the decision (Keystone 1996).

Finally, trust among partners is an important ingredient for success, one that takes conscious effort to achieve (Keystone 1996). Enabling trust to grow within a partnership can be achieved through good listening, membership diversity, and willingness to compromise.

Even though a partnership meets the criteria described above, certain challenges still must be overcome for the initiative to succeed. These include such aspects as compatibility of technical components (such as land classification systems, data, and analytical tools) (Haufler 1995), statutes and regulations, tracking landscape changes, modeling long-range futures, and balancing objectives.

Conserving biological diversity through partnerships across multiple ownerships and boundaries requires that landowners use compatible ecological land classification systems. Such classifications do not have to be identical but must have similar-enough properties that enable "cross-walking" among systems. In addition to using a compatible ecological land classification system, the partners in a collaborative process will need to identify the types of data and analysis tools needed to assess their mutually agreed upon goals and objectives (Haufler 1995). For example, if habitat suitability models are determined useful, then the models used by multiple partners must be compatible and the various partners may need to provide the data describing the species' habitat requirements.

A variety of federal and state statutes and regulations have the capability of challenging well-intentioned partnership efforts. These include, but are not limited to, the Federal Advisory Committee Act (FACA), Sherman Antitrust Act, Endangered Species Act (ESA), and federal land management agency planning acts (for example, the National Forest Management Act [NFMA] and National Environmental Policy Act [NEPA]). FACA can be a barrier to collaborative partnerships where federal agency personnel are involved in interjurisdictional coordination efforts with policy ramifications. Antitrust laws were enacted to prohibit agreements among competitors that reduce competition. Such laws impose legal risks to participants when two or more organizations producing a commodity are part of a coordinated planning process in which the potential outcome may be perceived as an agreement to cooperatively reduce output. The ESA has the potential to be a barrier to broad-based conservation of biological diversity because of costly and cumbersome mechanisms of the act. For example, constraints established for single-species protection may complicate broader efforts at maintenance of biological diversity. One example of the potential contradiction has been realized in the management of the northern spotted owl on the east side of the Cascades in the western United States. The ESA requires protection of the species, even in landscapes or smaller geographic areas where historical populations of the species may have been low or absent. Therefore, the presence and protection of a single species outside its historical range compromise conservation of biological diversity at a broader scale.

Finally, legislative actions that require specific planning processes by federal land management agencies can become a barrier to broader coordinated efforts because they inhibit federal involvement in such efforts. For example, legislation such as NFMA guides the activities of the national forests by detailing its responsibilities with regard to land management planning. If partnership efforts move away from the goals and objectives stated under NFMA, then participation by the Forest Service may become limited.

Long-range coordination of the conservation of biological diversity will continue to be challenged by the inability of land managers to accurately predict landscape changes and corresponding changes in biodiversity. The ability to model long-range futures of biodiversity is beginning to develop; however, the need to remain adaptive is critical for any collaborative process to be successful, given the relative uncertainties of current technologies and varied objectives. Finally, the challenges of balancing multiple objectives can be overcome only through the use of strong collaborative partnerships where ecological, social, economic, political, and legal interests are represented, trusted, and considered by all participants.

Public Relations

Conserving biological diversity by means of a partnership approach requires that stakeholders become involved at all levels and that strong public relations become a priority from the beginning. In creating a public relations campaign, the participants need to consider the tools that encourage effective communication and the means by which to gain support from and educate the right stakeholders at the right time.

Complex ecological theories, social uncertainties, and challenging economic positions must be communicated clearly and directly and yet take into account the integrity of the systems to be maintained. Any message must engage the public with respect to the importance of the issue. This may be accomplished by appealing to the innate desire to see nature persist or through education with regard to the issue. Communicating the potential trade-offs between social, economic, and ecological objectives is generally perceived to be up-front and unbiased; presenting only one side of an issue is generally seen as biased and having hidden agendas.

Using a variety of media is important when trying to educate a broad group of potentially interested parties. Traditionally, ecological concerns and management decisions were discussed within the scientific community through professional publications or "white papers." Today, other media sources must be used to reach all the constituents of a concern. These include television, radio, newspapers and magazines, public notices, public meetings, workshops, and educational materials for use in schools.

However, it is not enough to take the complex writings or thoughts out of professional publications and insert them into the Sunday paper. In addition to using a wide variety of media sources, proponents of an initiative need to consider how they will convey the message to the general audience of a particular media. For example, radio segments explaining the importance of a particular conservation effort should be short and to the point so as not to turn the listener off with unnecessary facts. On the other hand, if a public meeting has been called to discuss an issue, more detail can be given through the use of visual displays and question-answer periods in which the audience is engaged in the discussion.

The greatest challenge in any public relations effort promoting a biodiversity project is to educate and gain support from constituents. Any one issue may require support from technical specialists, political officials, general citizens, environmental advocates, government employees, the private sector, and financial institutions. Furthermore, support, both internal and external, often must be gained by the sponsoring organization or agency. The key to gaining this support is to know your audience. Understanding the history behind past and present relationships with a respective ally is necessary. Similarly, understanding external influences that may be acting on the situation is important to keep in mind when negotiating for support.

You also need to know when you have received support. Measures of "sign-on" may come in many forms, from financial support to written endorsement, "hand-shake agreements," to general support with limited commitment. Understanding the level of sign-on is important in order to spend time and money wisely. For example, if a prospective partner offers only general support with limited resources to back up his or her commitment, then your future efforts would be wisely spent in trying to increase his or her support or, if more practical, working to gain the support of another partner.

Educating the right stakeholders from the beginning is also critical in order to generate and maintain support. Working to educate the public on concepts and principles of biological diversity may allow topics to be discussed rather than debated. Furthermore, working to eliminate misunderstandings will often focus the topic and allow for support and success to be obtained more quickly.

Public relations efforts are becoming increasingly important in today's business environment, including the business of conserving biological diversity. Understanding human nature and the intricacies of communication are important tools to develop when entering into the public arena. Strong partnerships will last only if communication is functioning and appropriate support has been obtained.

Applying Adaptive Management Principles

Maintaining and enhancing biological diversity in today's environment under conditions of conflicting demands, dynamic landscapes, and a changing knowledge base demand that management be adaptive by design—that is, that it emphasize experimentation and change in response to new information (Haufler et al. 1996; Keystone 1996). Although a relatively new concept (Nudds, Chapter 11), adaptive resource management has emerged as a way of dealing with inherent uncertainties within and among ecological, social, and economic systems. Adaptive management addresses uncertainty by structuring initiatives as experiments in which the results are used to continually correct the course of action (Walters 1986; Keystone 1996; Trauger, Chapter 12).

Long-term monitoring is a critical component in managing by experimentation. Long-term monitoring programs must be well-thought-out experiments in themselves and maintain the ability to adapt in a manner similar to that which is being monitored. As described by Nudds (Chapter 11), adaptive management principles can be used to design cost-effective programs for inventory and monitoring.

Finally, a distinction should be made between active and passive adaptive resource management. Passive adaptive management is a less-refined, "trial-by-error" approach, whereas active adaptive management is a structured quantitative strategy (Keystone 1996). Efforts that are designed to ensure conservation of biodiversity should employ active adaptive management strategies to ensure a faster rate of learning and greater accountability of management goals. Requirements for successful adaptive management include (1) an understanding of system function and outputs, (2) establishment of quantified objectives and controls, (3) initiating a course of action, (4) monitoring and evaluating the outcomes, and (5) reviewing the goals and objectives and redirecting activities if necessary (Keystone 1996).

Adaptive management encourages active participation of all partners through routine checks on balancing ecological, social, and economic objectives. The inclusion of checks and balances in a process fosters trust among participants and ensures decision making by consensus rather than by directive.

Summary and Conclusions

The maintenance of biological diversity has become increasingly important in natural resource management in the recent decade. However, to date, adequate processes have not been developed. It is generally agreed that consideration must be given to ecological, political, economic, social, and legal concerns. Ecological objectives must be clear and other considerations (political, economic, social) applied within their context. Spatial and temporal scales are important to recognize and define. Adequate maintenance of biodiversity generally requires management decisions over a large geographic area. However, fine-resolution mapping and classification are necessary to fully understand the complexity of the landscape. Time frames beyond traditional planning horizons are often necessary to consider when implementing appropriate processes.

Balance of all appropriate concerns can be achieved only through partnerships with voluntary collaboration as the foundation. Ownership in any decision-making process is critical and can be achieved only through involvement by all interested participants from the outset. Challenges to such partnerships include compatibility of data and tools across multiple ownerships, legal constraints, and private property rights. However, such challenges can be overcome through the use of strong collaborative partnerships. Maintaining and enhancing biological diversity in today's changing environment can be achieved through the use of adaptive management principles. Long-term monitoring becomes a critical component of any initiative designed to be adaptive. The checks and balances that are realized through adaptive management planning enhance the success of any partnership.

Communicating the intent of maintaining biological diversity is a challenge that can be overcome only through the use of good public relations efforts. Conveying complex issues in a simple-to-understand format requires using a variety of media and understanding the audience. Finally, generating and maintaining support for specific initiatives are critical to the successful implementation of any process. Support must come from within as well as from outside. Understanding relationships among stakeholders and recognizing internal politics are critical to success of any public relations campaign.

Implementation of an effective process for the conservation of biological diversity are about understanding the ecology of the system within today's political environment. A successful process must be ecologically sound, based on partnerships, derived from adaptive management principles, and have the support of all interested parties. Conservation of biological diversity requires recognizing and balancing trade-offs while maintaining the complete range of ecological processes.

Chapter 16

Future Innovations
and Research Needs

Henry Campa III,
Richard K. Baydack, and
Jonathan B. Haufler

All the approaches discussed in this book are hypotheses about how to conserve biological diversity and, as such, need to be evaluated. There are also numerous research topics that need to be studied through application of these approaches so we can be more effective at conserving biological diversity. This chapter discusses research needs and potential management changes that should be addressed now and in the future to conserve biodiversity. More specifically, we discuss the need for better access to historical ecological data and an understanding of the historical range of variability (HRV) of ecosystem characteristics. Also, better integration of geographic information systems (GIS) technology with more sophisticated models of species habitat requirements is required to evaluate current and potential impacts of management practices on landscape composition and structure, the availability of ecological resources, habitat quality and quality, and the viability of species populations. Such tools and models will have to be flexible analytical techniques for evaluating the effects of management practices on the conservation of biological diversity among multiple scales of time and space.

Integrating Terrestrial, Wetland, and Aquatic Ecosystem Management

Currently, most natural resources planning and management goes on in distinct ecosystems (terrestrial, wetland, aquatic), often within different landscapes (counties, districts, sections, regions of states, tracts of land within an ownership, watersheds, ecoregions). Nevertheless, how one type of ecosystem is managed may have impacts on the biological diversity of adjacent ecosystems. Moreover, because ecosystem processes are a part of biological diversity conservation, natural resource managers must also consider the effects that management practices in one ecosystem type will have on the ecological process of other types.

Making planning and management decisions for separate ecosystems without considering other ecosystem types can also be problematic for wildlife species that use multiple types of ecosystems to meet their habitat requirements and at scales beyond the planning landscape or land ownership. For example, land-use practices in different types of forest and grassland ecosystems will produce different sediment yields (movement of soil), with some grasslands yielding approximately ten times more sediment than forests that are not harvested. But where does this soil go? How does soil loss affect the productivity, species composition, and structure of the ecosystem from which it is being lost as well as where it will be deposited? Additionally, how will changes in these ecosystem characteristics affect the biological diversity of each individual ecosystem?

Because of the uncertainty of the answers to these types of questions, we encourage natural resource managers to develop integrated, landscape-level research projects that investigate the ecological effects that manipulations in individual ecosystems have on the composition, structure, and processes of all other associated ecosystems. Additionally, assessing the biological diversity in response to such manipulations within each ecosystem-type may give managers valuable insights into how manipulating one ecosystem can affect the biodiversity and integrity of other ecosystems.

Answering these kinds of questions will require that all natural resource professionals use a more integrated approach in developing research projects and in implementing management plans based on their findings. This paradigm is a dramatic change from what has been done historically. Aquatic and terrestrial managers will need more information and planning before initiating management activities, especially in terms of the type of manipulation and its timing, scale, and juxtaposition to other disturbances.

Integrating Ecological Information at Different Scales

The element of scale (that is, space, time, and levels of biological organization) is a critical component to consider in biodiversity conservation. Unfortunately, natural resource managers have only recently started investigating the implications of scale when planning and conducting management activities. For example, we are just beginning to ask if small-scale habitat manipulations have a long-term impact on the functioning of larger geographical areas? Or do habitat manipulations have to be larger and more intense to conserve the biological diversity associated with some types of ecosystems?

Natural resource managers need to address the conservation status and needs of all elements of biodiversity using landscape-level ecological boundaries rather than other approaches. In addition, because different elements of biodiversity use landscapes of different scales, it is important that natural resource managers attempting to conserve biodiversity can make management decisions at a broad scale that will benefit species of conservation concern. For example, conserving a metapopulation throughout a relatively broad landscape may require that natural resource managers gain a greater understanding of the roles habitat patches within the planning landscape play in conserving the species. We believe that to develop these capabilities, natural resource managers need to investigate the relationships between planning and managing on a broad scale and responses of biological diversity and ecosystem integrity on smaller ecological scales, and vice versa.

Compatible Land Classification Systems That Incorporate Ecological Classification Systems

In the future, practices to conserve biological diversity will be facilitated by developing, experimenting with, and implementing land classification systems that incorporate ecological classification systems (ECS). These classification systems should have two general characteristics: they should be compatible among management agencies, organizations, and industries; and they should be based on potential vegetation or site conditions as well as current dominant vegetation. Using compatible classification systems will allow different types of natural resource managers to work cooperatively to maintain biological diversity at appropriate spatial scales. Because the vegetation associated with ecosystems is influenced by past management activities and natural disturbances, using existing vegetation as the basis for a classification system will be problematic if it will be used to plan future manipulations to conserve biodiversity (Roloff and Haufler 1993). Therefore, classification systems should account for vegetation change and be based on all the ecological parameters that have the greatest influence on the potential vegetation associated with ecological units.

Applications of Compatible Geographic Information System Technology

For natural resource managers to be effective at conserving biological diversity across diverse landscapes, they will require larger, more complex data sets. Use of compatible, user-friendly GIS technology will allow managers to use these data to quantify baseline ecological conditions among ownerships (Neraasen and Nelson, Chapter 9), assess how the distribution of various elements of biological diversity change over time (Haufler et al. Chapter 7), assess how biodiversity is altered by manipulations over various scales, and integrate various data sets.

To facilitate the use of new GIS technology for these applications, managers will need to integrate it with global positioning systems technology and remote sensing. This integration will provide managers with access to a greater number of data sources and control over GIS databases. As part of this process, the databases that are used in association with GIS technology will need to be validated if they are to be an integral part in driving natural resources planning and management decisions. Therefore, the databases must be evaluated for both their needed resolution to address specific biodiversity issues and their accuracy to ensure that mapped information reflects actual landscape conditions.

Sources of Ecological Data

Landscape-level planning and management for biodiversity conservation have often been hindered by the lack of comprehensive, compatible, high-quality information about the ecological resources we are trying to manage (Scott et al., Chapter 4). For example, when natural resource managers wish to manage ecosystems within historic range of variability (HRV) or some socially acceptable limit on the basis of the historic range, what will be the sources of historical information and is it available for broad landscapes or just local areas? Aplet and Keeton (see Chapter 5) suggested additional research on tree rings, fire scars, and pollen cores is useful and may continue to be in providing important sources of historical data. Specifically, these sources of data may give us a better understanding of the past disturbance regimes associated with various habitat types and their influence on vegetation species composition, structure, and spatial distribution as well as the species associated with the successional stages produced by disturbances.

Application of Historical Range of Variability

Conservation of biological diversity may be enhanced with a better understanding of HRV and the quantities and distribution of successional stages. What the limits of ecosystem characteristics are in response to natural disturbances is a crucial piece of information if we wish to prescribe management practices that to some degree will emulate these fluctuating characteristics.

Moreover, natural resource managers need to be cautious when making management decisions at the stand level while ignoring their potential ecological effects within the larger landscape. These decisions may impair the temporal heterogeneity necessary to maintain naturally functioning ecosystems following past cumulative human-induced disturbances (Aplet and Keeton, Chapter 5). To what extent these types of human disturbances have altered the dynamic nature of ecosystems is a future research challenge.

Population Viability Analysis and Linkages with Habitat Analysis and Management

One critical component in conserving biodiversity is that populations have to be maintained at viable numbers. Therefore, determining the effective population size of critically important species (threatened and endangered species, indicator species for fine-filter assessments) would be useful in setting minimum population management goals. Eventually our ability to quantify effective population sizes should be linked with a coarse-filter level of analysis (Haufler et al. Chapter 7) to facilitate our ability to make more effective population and habitat management decisions on a landscape scale.

Collecting the necessary demographic data (the number of male and female breeders, number of young born to females per breeding interval, the probability of males and female young surviving to the mean age of reproduction and breeding, and so forth) to calculate an effective population size of a species is often problematic because of the length of time it takes to collect these data and the cost associated with such studies. Potentially, by the time an adequate data set is gathered, the population dynamics of a species has changed along with its habitat conditions. Additionally, specific assumptions need to be evaluated for a population viability analysis, namely, whether there is an equal sex ratio, random mating, equal probability for all individuals to produce young, nonoverlapping generations, and a relatively constant population. Ultimately, if natural resource managers can quantify an effective population size for a species that is based on valid assumptions, then maintaining populations above this level will presumably maintain the genetic variability within a species that allows for its long-term persistence. This relationship—maintenance of an effective population size that is genetically diverse and that will persist over time—needs further evaluation.

A second area for additional research is the link between the results of population viability analysis with habitat quality and quantity. This may best be accomplished through a coarse-filter approach because it allows planning for a mix of ecosystems on a landscape scale (Haufler et al. Chapter 7). If this linkage is desirable, natural resource managers should evaluate the effectiveness of the coarse-filter analyses to assess ecological conditions throughout the planning landscape.

Evaluating Habitat Management Techniques That Emulate Natural Disturbances

Runkle (1985:31) asked the question, "Do the species otherwise adapted to an area ... determine the disturbance regime, or does the potential disturbance regime in an area determine the vegetation?" He then added that "[b]oth factors may interact simultaneously and reciprocally, so that simple causation is impossible to detect." For resource managers, investigating these relationships may be critical to the development of new management techniques that emulate natural disturbances so that we can create the ecological conditions necessary to conserve biological diversity while simultaneously using natural resources to meet economic and recreational demands.

In some instances, however, ecological systems may be so altered by humans that initiating management practices that emulate natural disturbances would cause deleterious impacts or they may not be socially acceptable. For example, we need to consider whether remnant prairies in the eastern United States can be burned to restore biodiversity when they have not been manipulated in numerous years and especially when they are adjacent to farms and homes. But then, will the contributions these ecosystems make to biodiversity conservation be lost unless succession is set back by other unexplored management practices? The development, experimentation, and implementation of habitat management practices that may emulate the natural disturbances associated with ecosystems are encouraged; however, these practices must be realistic and done in an adaptive management approach (Nudds, Chapter 11).

Controlling Exotic Species

Exotic species have changed the composition, structure, and ecological processes of various types of ecosystems (Campa and Hanaburgh, Chapter 13). To what degree the implementation of management regimes that emulate natural disturbances can be used to restore ecosystems degraded by exotics or to minimize future invasions are topics for future research. Scientific experiments that evaluate the effects of management practices on the characteristics of ecosystems and the response of exotics in invaded and control sites may provide insights into reducing their threats to biodiversity. Limiting factors for implementing management practices that control exotic species undoubtedly will be the costs and the degree to which an ecosystem has been altered by exotics.

Applying Fine-Filter Strategies to Biodiversity Conservation

While the need to maintain variety of life at all levels of biological organization as well as ecological processes is essential for biodiversity conservation, historically most natural resources research, planning, and management has been done on or for individual species.

Therefore, a problem facing natural resources managers is how they can determine whether the goal of maintaining biodiversity is being met if most of their activities are focusing on individual species. One response to this dilemma has been to apply fine-filter strategies (Wall, Chapter 8). Some researchers (for example, Landres et al. 1988; Niemi et al. 1997) have scrutinized the use of some of these approaches.

To more effectively use the fine-filter strategy for biodiversity conservation, further knowledge is needed on such topics as how population attributes of a cohort of species are representative of changes in the population status or habitat quality of a diversity of species. In addition, if guilds are used to assess the effectiveness of management practices in conserving biodiversity, the degree of similarity of guild members may indicate their effectiveness in assessing biodiversity conservation. And, lastly, the use of fine-filter strategies requires that natural resource managers initially have the best scientific data available about the species or cohorts of species of interest. Gaining a greater understanding of the life history of species as well as basic information on the composition, distribution, and relative abundance of all types of species (plants, vertebrates, and invertebrates) is critical to the selection of indicator species and constructing guilds. Presently, many of these types of data are incomplete, have not been validated, or do not exist (Trauger, Chapter 12).

The Future Role of Education in Biodiversity Conservation

Throughout this volume, scientists have shared their thoughts on approaches to conserving biodiversity and challenges in meeting this objective. Most readily admit that their approaches are experimental. If we are to be effective stewards of biodiversity in the future, we must continue our conversations about how these experiments are working. In essence, this may require that we use new educational formats to inform ourselves and others about the effectiveness and complexities of current and future approaches to biodiversity conservation. For example, professional meetings frequently include formal presentation sessions where researchers present their best science on a given topic. These presentations are informative and illustrated with the latest technology in computer graphics. However, at the end of twenty minutes, speakers step down and often leave the session. How much exchange of information has taken place under this scenario? Do speakers hear other presentations on different perspectives of the same topic? In this format, speakers disseminate their information but frequently do not have time to answer, discuss complex topics, or ask questions. At some conferences, sessions are followed by a "panel discussion," which are often very informative because scientists have additional time to comment on and discuss the science they presented. But is this the best format to use in the future for educating others and keeping ourselves current on the details of such complex topics as approaches to conserving biodiversity?

Perhaps because scientists are now experimenting with approaches to conserve biodiversity we are at a crossroads. We are no longer just talking about what biodiversity is, what it isn't, and whether we can manage for it. As we gain a better understanding of ecological relationships and the effectiveness of the approaches we are now using, we should look at educational opportunities and formats for staying current on field experimentation and share these results with colleagues. To accomplish these educational objectives, perhaps we must look beyond conducting the typical symposium to discuss biodiversity strategies and consider conducting "workshops" both within and among professional societies, nongovernment organizations, industry, and groups of private landowners. In workshops, natural resource professionals could share the experiences they have had in working with various strategies, make suggestions about how other natural resource professionals could develop or adopt approaches to meet their specific needs, and help others develop methodologies to test the effectiveness of strategies.

If we are going to be effective in conserving biodiversity, we need to interact with a broader audience than the usual people we talk to on a day-to-day basis—in essence, do a better job of disseminating information. The future of biodiversity conservation depends on developing educational programs and processes within agencies, organizations, and professional societies that show what can be done and how approaches can be demonstrated or implemented. A major goal for future educational programs on biodiversity conservation may be to provide diverse groups of natural resource managers with guidance on using the approaches discussed in this volume.

Summary and Conclusions

The practical approaches to conserve biodiversity discussed in this book will be most effective if they are evaluated in applied projects under adaptive management principles. Additionally, as with any complex subject, many research topics need to be addressed so that these approaches can be made more effective and new techniques or innovations developed. The research areas to address the complex topic of biodiversity conservation are also complex. As some people say, "If it were easy, others would have already done it." We propose that additional research should span all scales of time, space, and levels of biological organization as well as provide us with a better understanding of ecological processes. Natural resource managers should also continue to look at new, innovative ways to integrate current and future data. This integration of data undoubtedly will require the use of sophisticated computer hardware and software. And, lastly, if we are going to be successful in meeting our goal of biodiversity conservation, we must continue our conversations about the subject in such forums as the symposium that spurred the idea to develop this volume. Becoming better stewards of biodiversity will require us to conduct rigorous research as well as to share our successes and failures as we strive toward this goal.

Closing Perspectives

Richard K. Baydack

To conserve biological diversity or not to conserve biological diversity—that is no longer the question. Whether it is nobler to use a bioreserve strategy, an emphasis-area approach, coarse- or fine-filter strategies, or some combination of these remains to be determined, especially for each individual management scenario. Or perchance to dream, that some newfound approach may yet emerge that betters the rest, that meets the test and becomes the best. That in fact may be the important development of the next decade.

With apologies to William Shakespeare, the above paragraph summarizes quite succinctly our overall goal and purpose and, we hope, the outcome of this book in your mind. We have tried to convey that the future of our planet depends on the human race "getting its act together" to conserve the variety of life and its processes on which all flora and fauna depend. We have similarly tried to provide a rationale for what this "variety" means in the technical sense and how it can be defined. We have enlisted the assistance of a number of our colleagues to help document the ways in which biological life-forms and their processes have been successfully conserved. These approaches have been described in situations where the strategies have been shown to work, and we anticipate their application can be extrapolated to a wider universe beyond. Of likely most importance, we have limited our descriptions and discussions to practical techniques—those that can and must be used by natural resource managers to maintain the quality of life to which residents of Earth have become accustomed. All of these good efforts are not without complication, and we have tried to examine those difficulties by providing alternatives that managers can and should explore in the years ahead. To this end research will be critical, and we have provided a prioritized listing of the most important issues that will require the attention of our best scientists.

But for today, managers must continue to manage, and therefore we have offered a set of guidelines to make the job easier and more efficient. In short, this book has a little bit of everything for everyone in the natural resource professions. We hope that you examine it when you can, employ it where you can, and improve upon it as much as you can.

And finally, to those of you who have made your way through these pages, please keep in mind this basic idea: Because biological diversity is perhaps "everything," the importance of the concept cannot be overstated. Its grandeur is monumental, its impact is all-encompassing, and its continuity is forever. We ask that you guard against its loss and that you work to pass on its blessings to your next generation. If we succeed in doing this much, we can rest easy knowing that our work was not in vain.

References

Chapter 1

Anderson, W. L., and R. W. Storer. 1976. Factors influencing Kirtland's warbler nesting success. *The Jack Pine Warbler* 54: 105–115.

Council on Environmental Quality. 1981. *The Global 2000 report to the President*. Vol. II. Washington, D.C.: Council on Environmental Quality and the U.S. Department of State.

DeLong, D. C., Jr. 1996. Defining biodiversity. *Wildlife Society Bulletin* 24: 738–749.

Eckholm, E. 1975. *The other energy crisis: Firewood*. Worldwatch Paper 1. Washington, D.C.: Worldwatch Institute.

Environment Canada. 1996. *The Canadian biodiversity strategy*. Ottawa: Environment Canada.

Erickson, J. R., and D. W. Toweill. 1994. Forest health and wildlife habitat management on the Boise National Forest. *Idaho Journal of Sustainable Forestry* 2: 389–409.

Farnsworth, N. R., and R. W. Morris. 1976. Higher plants—The sleeping giant of drug development. *American Journal of Pharmaceutical Education* 148: 46–52.

Grumbine, R. E. 1994. What is ecosystem management? *Conservation Biology* 8: 27–38.

Hanaburgh, C. 1995. Wildlife use of native and introduced grasslands in Michigan. M.S. thesis, Michigan State University.

Haufler, J. B., C. A. Mehl, and G. J. Roloff. 1996. Using a coarse-filter approach with species assessment for ecosystem management. *Wildlife Society Bulletin* 24: 200–208.

Heywood, V. E., ed. 1995. *Global biodiversity assessment*. United Nations Environment Programme report. New York: Cambridge University Press.

Hunter, M. L., Jr. 1990. *Wildlife, forests, and forestry principles of managing forests for biological diversity*. Englewood Cliffs, N.J.: Prentice Hall.

Johnson, S. P., ed. 1993. *The Earth Summit: The United Nations conference on environment and development*. Bk. 1. Norwell, Mass.: Kluwer Academic Publishers.

Kaufmann, M. R., R. T. Graham, D. A. Boyce, Jr., W. H. Moir, L. Perry, R. T. Reynolds, R. Bassett, P. Mehlhop, C. B. Edminster, W. M. Block, and

P. S. Corn. 1994. *An ecological basis for ecosystem management.* USDA Forest Service, General Technical Report RM-246.

Keystone Center, The. 1991. *Final consensus report of the Keystone Policy Dialogue on biological diversity on federal lands.* Keystone, Colo.: The Keystone Center.

Line, L. 1964. The bird worth a forest fire. *Audubon Magazine* (November–December): 16–17.

Lovejoy, T. E. 1997. Biodiversity: What is it? In *Biodiversity II: Understanding and protecting our biological resources,* ed. M. L. Reaka-Kudla, D. E. Wilson, and E. O. Wilson, 7–14. Washington, D.C.: Joseph Henry Press.

McNaughton, S. J. 1975. Structure and function of Serengeti grasslands. *Annual report of the Serengeti Research Institute 1974–1975*: 51–61.

McNeely, J. A., K. R. Miller, W. V. Reid, R. A. Mittermeier, and T. B. Werner. 1990. *Conserving the world's biological diversity.* Gland, Switzerland: International Union for Conservation of Nature and Natural Resources; Washington, D.C.: World Resources Institute, Conservation International, World Wildlife Fund—U.S., and the World Bank.

Meffe, G. K., and C. R. Carroll. 1994. *Principles of conservation biology.* Sunderland, Mass.: Sinauer Associates.

Miller, K. R., J. Furtado, C. DeKlemm, J. A. McNeely, N. Myers, M. E. Soulé, and M. C. Trexler. 1985. Issues on the preservation of biological diversity. In *The global possible: Resources, development, and the new century,* ed. R. Repetto, 337–362. New Haven: Yale University Press.

Muthee, L. W. 1991. Ecological impacts of tourist use on habitats and pressure-point animal species. In *Tourist attitudes and use impacts in Maasai Mara National Reserve,* ed. C. G. Gakahu, 18–38. Nairobi, Kenya: Wildlife Conservation International.

Norse, E. A., and R. E. McManus. 1980. Ecology and living resources biological diversity. In *Council on Environmental Quality, Eleventh Annual Report,* 31–80. Washington, D.C.: Council of Environmental Quality.

Noss, R. F., and A. T. Cooperrider. 1994. *Saving nature's legacy—Protecting and restoring biodiversity.* Washington, D.C.: Island Press.

Office of Technology Assessment (OTA). 1987. *Technologies to maintain biological diversity.* OTA-F-330. Washington, D.C.: Government Printing Office.

Packer, C., A. E. Pusey, H. Rowley, D. A. Gilbert, J. Martenson, and S. J. O'Brien. 1991. Case study of a population bottleneck: Lions of the Ngorongoro Crater. *Conservation Biology* 5: 219–230.

Pineda, F. D., M. A. Cadao, J. M. de Miguel, and J. Montsalvo, eds. 1991. *Biological diversity.* Madrid: Fundacion Ramon Areces.

Reaka-Kudla, M. J., D. E. Wilson, and E. O. Wilson, eds. 1997. *Biodiversity II: Understanding and protecting our biological resources.* Washington, D.C.: Joseph Henry Press.

Reid, W. V., and K. R. Miller. 1989. *Keeping options alive: The scientific basis for conserving biodiversity.* Washington, D.C.: World Resources Institute.

Ryan, J. C. 1992. Conserving biological diversity. In *State of the world 1992,* ed. L. R. Brown, C. Flavin, S. Postel, and L. Starke, 9–26. New York: W. W. Norton.

Salwasser, H. 1990. Conserving biological diversity: A perspective on scope and approaches. *Forest Ecology and Management* 35: 75–90.

Skiba, G. 1994. Are definitions really important? *Biological Diversity Working Group Newsletter* (May): 3.

Terborgh, J. W. 1986. Keystone plant resources in the tropical forest. In *Conservation biology: The science of scarcity and diversity*, ed. M. E. Soulé, 330–344. Sunderland, Mass.: Sinauer Associates.

United Nations Environment Programme. 1991. *Fourth revised draft convention on biological diversity.*

U.S. Congress. 1991. *National Biological Diversity and Environmental Research Act of 1991.* 102d Cong., 1st sess., S. 58, H.R. 585.

U.S. Fish and Wildlife Service. 1992. *Conserving biological diversity—A strategy to maintain biodiversity and ecosystem function in the Northeast.* Newton Corner, Mass.: U.S. Fish and Wildlife Service, Region 5.

U.S. Forest Service. 1990. *Conserving our heritage—America's biodiversity.* Washington, D.C.: U.S. Forest Service.

Wilson, E. O. 1992. *The diversity of life.* New York: W. W. Norton.

_____. 1997. Introduction to *Biodiversity II: Understanding and protecting our biological resources*, ed. M. J. Reaka-Kudla, D. E. Wilson, and E. O. Wilson. Washington, D.C.: Joseph Henry Press.

Wilson, E. O., and F. M. Peter, eds. 1988. *Biodiversity.* Washington, D.C.: National Academy Press.

Chapter 2

Alverson, W. S., W. Kuhlmann, and D. M. Waller. 1994. *Wild forests: Conservation biology and public policy.* Washington, D.C.: Island Press.

Bailey, R. G. 1995. *Descriptions of the ecoregions of the United States.* USDA Forest Service, Misc. Publ. 1391.

_____. 1996. Ecosystem geography. New York: Springer-Verlag.

Blockstein, D. E. 1995. A strategic approach for biodiversity conservation. *Wildlife Society Bulletin* 23: 365–369.

Daubenmire, R. F. 1968. *Plant communities: A textbook of plant synecology.* New York: Harper and Row.

DellaSala, D. A., D. M. Olson, S. E. Barth, S. L. Crane, and S. A. Primm. 1995. Forest health: Moving beyond rhetoric to restore healthy landscapes in the inland Northwest. *Wildlife Society Bulletin* 23: 346–356.

DellaSala, D. A., J. R. Strittholt, R. F. Noss, and D. M. Olson. 1996. A critical role for core reserves in managing inland northwest landscapes for natural resources and biodiversity. *Wildlife Society Bulletin* 24: 209–221.

Ecomap. 1993. *National hierarchical framework of ecological units.* Washington, D.C.: USDA Forest Service.

Everett, R. L., P. F. Hessburg, and T. R. Lillybridge. 1994. Emphasis areas as an alternative to buffer zones and reserved areas in the conservation of biodiversity and ecosystem processes. *Journal of Sustainable Forestry* 2: 283–292.

Everett, R. F., and J. Lehmkuhl. 1996. An emphasis-use approach to conserving biodiversity. *Wildlife Society Bulletin* 24: 192–199.

FEMAT. 1993. *Draft supplemental environmental impact statement on management of habitat for late-successional and old-growth forest related species within the range of the northern spotted owl.* Washington, D.C.: USDA Forest Service.

Grumbine, R. E. 1994. What is ecosystem management? *Conservation Biology* 8: 27–38.

Hall, L. S., P. R. Krausman, and M. L. Morrison. 1997. The habitat concept and a plea for standard terminology. *Wildlife Society Bulletin* 25: 173–182.

Harris, L. D. 1984. *The fragmented forest.* Chicago: University of Chicago Press.

Haufler, J. B., T. R. Crow, and D. S. Wilcove. In press. Scale considerations for ecosystem management. *Proceedings of the Ecological Stewardship Workshop.*

Haufler, J. B., C. A. Mehl, and G. J. Roloff. 1996. Using a coarse-filter approach with species assessment for ecosystem management. *Wildlife Society Bulletin* 24: 200–208.

Holling, C. S., ed. 1978. *Adaptive environmental assessment and management.* New York: John Wiley and Sons.

Honadle, G. 1993. Institutional constraints on sustainable resource use: Lessons from the tropics showing that resource overexploitation is not just an attitude problem and conservation education is not enough. In *Defining sustainable forestry,* ed. G. H. Aplet, N. Johnson, J. T. Olson, and V. A. Sample, 90–122. Washington, D. C.: Island Press.

Hoover, R. L., and D. L. Wills. 1984. *Managing forested lands for wildlife.* Denver: Colorado Division of Wildlife.

Huff, M., K. Mellen, and R. Hagestedt. 1994. Interpreting landscape patterns: A vertebrate habitat relationships approach. *Analysis Notes* 4(2): 6. USDA Forest Service, WO-Ecosystem Management Analysis Center, Washington, D.C.

Hunter, M. L. 1990. *Wildlife, forests, and forestry: Principles of managing forests for biodiversity.* Englewood Cliffs, N.J.: Prentice Hall.

Kaufmann, M. R., R. T. Graham, D. A. Boyce, Jr., W. H. Moir, L. Perry, R. T. Reynolds, R. L. Bassett, P. Mehlhop, C. B. Edminster, W. M. Block, and P. S. Corn. 1994. *An ecological basis for ecosystem management.* USDA Forest Service, General Technical Report RM-246.

Marcot, B. G., M. J. Wisdon, H. W. Li, and G. C. Castillo. 1994. *Managing for featured, threatened, endangered, and sensitive species and unique habitats for ecosystem sustainability.* USDA Forest Service, General Technical Report PNW-GTR-329.

Maxwell, J. R., C. J. Edwards, M. E. Jensen, S. J. Paustian, H. Parrott, and D. M. Hill. 1995. *A hierarchical framework of aquatic ecological units in North America (Nearctic zone).* USDA Forest Service, General Technical Report NC-176.

McMahon, J. 1993. *An ecological reserves system for Maine: Benchmarks in a changing landscape.* Bangor: Natural Resources Policy Division, Maine State Planning Office.

Morgan, P., G. H. Aplet, J. B. Haufler, H. C. Humphries, M. M. Moore, and W. D. Wilson. 1994. Historical range of variability: A useful tool for evaluating ecosystem change. *Journal of Sustainable Forestry* 2: 87–112.

Noss, R. F. 1994. The wildlands project: Land conservation strategy. *Wild Earth, The Wildlands Project* (special issue): 10–25.

————. 1996. On attacking a caricature of resources: response to Everett and Lehmkuhl. *Wildlife Society Bulletin.* 24: 777–779.

Noss, R. F., and A. Y. Cooperrider. 1994. *Saving nature's legacy: Protecting and restoring biodiversity.* Washington, D.C.: Island Press.

Oliver, C. D. 1992. A landscape approach: Achieving and maintaining biodiversity and economic productivity. *Journal of Forestry* 90(9): 20–25.

————. 1994. Rebuilding biological diversity at the landscape level. *Proceedings of the Conference on forest health and fire danger in inland western forests.* Harmon Press.

Risbrudt, C. 1992. Sustaining ecological systems in the Northern Region. In *Taking an ecological approach to management,* 27–39. USDA Forest Service, Watershed and Air Management, WO-WSA-3.

Scott, J. M., F. Davis, B. Csuti, R. Noss, B. Butterfield, C. Groves, H. Anderson, S. Caicco, F. D'Erchia, T. Edwards, Jr., J. Ulliman, and R. G. Wright. 1993. Gap analysis: A geographic approach to protection of biological diversity. *Wildlife Monographs* 123: 1–41.

Thomas, J. W., tech. ed. 1979. Wildlife habitats in managed forests: The Blue Mountains of Oregon and Washington. *Agricultural Handbook*, no. 553. Washington, D.C.: USDA Forest Service.

USDA Forest Service. 1994a. *Reinvention of the Forest Service: The changes begin.* Washington, D.C.: USDA Forest Service.

————. 1994b. *Deadwood landscape analysis.* Boise: USDA Forest Service, Lowman Ranger District, Boise National Forest.

Walters, C. J., and C. S. Holling. 1990. Large-scale management experiments and learning by doing. *Ecology* 71: 2060–2068.

Chapter 3

Angle-Franzini, M. 1996. Unity of action, saving the redwoods. *Proceedings of the Conference on Coast Redwood Forest Ecology and Management,* June 18–20, 1996, ed. J. L. LeBlanc, 4–7. Arcata, Calif.: Humboldt State University.

Baker, E. 1984. *An island called California—An ecological introduction to its natural communities.* Berkeley: University of California Press.

Becking, R. W. 1982. *A pocket flora of the redwood forest.* Washington, D.C.: Island Press.

Burkhardt, H. 1994. *Maximizing forest productivity—Resource depletion and a strategy to resolve the crisis.* Ukiah, Calif.: Mendocino Environmental Center.

California Board of Forestry. 1990. *Recommendations to the California State Board of Forestry on the management of wildlife habitats under the Forest Practices Act.* Sacramento: The Wildlife Habitat/Forest Practice Task Force.

California Department of Forestry and Fire Protection. 1988. *California's forests and rangelands: Growing conflict over changing uses.* Sacramento: Forest and Rangeland Resources Assessment Program (FRRAP), California Department of Forestry and Fire Protection.

California Water Quality Control Board. 1995. *Adopting the 1996 regional water quality assessment, 303(d) list and prioritization update.* Santa Rosa, Calif.: California Regional Water Quality Control Board, North Coast Region.

Carlson, S. A., L. Fox III, and R. L. Garrett. 1994. Virtual GIS and ecosystem assessment in the Klamath Province, California–Oregon. ASPRS Technical Papers: GIS/LIS, Annual Convention, American Society of Photogrammetry and Remote Sensing, Bethesda, MD, pp. 133–141.

Christensen, N. L., A. M. Bartuska, J. H. Brown, S. Carpenter, C. D'Antonio, R. Francis, J. F. Franklin, J. A. MacMahon, R. F. Noss, D. J. Parsons, C. H. Peterson, M. G. Turner, and R. G. Woodmansee. 1996. The report of the Ecological Society of America committee on the scientific basis for ecosystem management. *Ecological Applications* 6(3): 665–691.

Clark, T. W., and D. Zaunbrecher. 1987. The Greater Yellowstone Ecosystem: The ecosystem concept in natural resource policy and management. *Renewable Resources Journal* 5(3): 8–16.

Cooperrider, A., L. Fox III, R. Garrett, and T. Hobbs. In press. Data collection, management, and inventory. In *Ecological stewardship: A common reference for ecosystem management,* Volume 2, eds. N. C. Johnson, A. J. Malk, W. T. Sexton, and R. Szaro. New York: Elsevier Science.

Crumpacker, D. W., S. W Hodge, D. F. Friedley, and W. P Gregg Jr. 1988. A preliminary assessment of the status of major terrestrial and wetland ecosystems on federal and Indian lands in the United States. *Conservation Biology* 2(1): 103–105.

Dawson, T. E. 1996. The use of fog precipitation by plants in coastal redwood forests. *Proceedings of the Conference on Coast Redwood Forest Ecology and Management,* June 18–20, 1996, ed. J. L. LeBlanc, 90–93. Arcata, Calif.: Humboldt State University.

Fox, L. III. 1996. Current status and distribution of coast redwood. *Proceedings of the Conference on Coast Redwood Forest Ecology and Management,* June 18–20, 1996, ed. J. L. LeBlanc, 18–20. Arcata, Calif.: Humboldt State University.

Jensen, D. B., M. Torn, and J. Harte. 1990. *In our own hands: A strategy for conserving biological diversity in California.* California Policy Seminar, research report. Berkeley: University of California.

Johns, D., and M. Soulé. 1996. Getting from here to there. *Wild Earth* 5(4): 32–36.

Jones and Stokes Associates, Inc. 1989. *Effects of timber management under the California Forest Practices Act on wildlife species and habitats: An eval-*

uation of selected issues. Report prepared for California Department of Forestry and Fire Protection. Sacramento: Jones and Stokes Associates.

Keystone Center, The. 1991. *Final consensus report of the Keystone Policy Dialogue on biological diversity on federal lands.* Keystone, Colo.: The Keystone Center.

Kuchler, A. W. 1964. Potential natural vegetation of the conterminous United States (map and manual). *American Geographic Society Special Publication 36.*

Leopold, A. 1941. Wilderness as a land laboratory. *Living Wilderness* 6: 3.

_____. 1949. *A Sand County almanac.* New York: Oxford University Press.

Leydet, F. 1969. *The last redwoods and the parkland of Redwood Creek.* San Francisco: Sierra Club Books.

LSA Associates, Inc. 1989. *Final report: Forest practice rules recommendations/wildlife.* Report prepared for California Department of Forestry and Fire Protection. Point Richmond, Calif.: LSA Association.

Maser, C. 1990. On the "naturalness" of natural areas: A perspective for the future. *National Areas Journal* 10(3): 129–133.

McNeely, J. A., K. R. Miller, W. V. Reid, R. A. Mittermeier, and T. B. Werner. 1990. *Conserving the world's biological diversity.* Gland, Switzerland: International Union for Conservation of Nature and Natural Resources; Washington, D.C.: World Resources Institute, Conservation International, World Wildlife Fund—U.S., and the World Bank.

Meffe, G. K., and C. R. Carroll. 1994. *Principles of conservation biology.* Sunderland, Mass.: Sinauer Associates.

Mount, J. F. 1995. *California rivers and streams—The conflict between fluvial process and land use.* Berkeley: University of California Press.

Moyle, P. B. 1994. The decline of anadromous fishes in California. *Conservation Biology* 8(3): 869–870.

National Marine Fisheries Service. 1996. Threatened status for central California coast coho salmon evolutionarily significant unit (ESU). *Federal Register* 61(212): 56138–56149.

_____. 1997a. Threatened status for southern Oregon/northern California coast evolutionarily significant unit (ESU) of coho salmon. *Federal Register* 62(87): 24588–24609.

_____. 1997b. Press release, August 11, 1997. Washington, D.C.

Newmark, W. D. 1985. Legal and biotic boundaries of western North American national parks: A problem of congruence. *Biological Conservation* 33: 197–208.

Noss, R. F. 1983. A regional landscape approach to maintain diversity. *BioScience* 33: 700–706.

_____. 1992. The Wildlands Project: Land conservation strategy. *Wild Earth* (special issue): 10–25.

_____. 1993. A bioregional conservation plan for the Oregon Coast Range. *Natural Areas Journal* 13: 276–290.

Noss, R. F., and A. Y. Cooperrider. 1994. *Saving nature's legacy—Protecting and restoring biodiversity.* Washington, D.C.: Island Press.

Noss, R. F., and L. D. Harris. 1986. Nodes, networks, and MUMs:

Preserving diversity at all scales. *Environmental Management* 10: 299–309.

Pace, F. 1991. The Klamath Corridors: Preserving biodiversity in the Klamath National Forest. In *Landscape linkage and biodiversity*, ed. W. Hudson, 105–116. Washington, D.C.: Island Press.

Primack, R. B. 1993. *Essentials of conservation biology.* Sunderland, Mass.: Sinauer Associates.

Reid, W. V., and K. R. Miller. 1989. *Keeping options alive: The scientific basis for conserving biodiversity.* Washington, D.C.: World Resources Institute.

Shafer, C. L. 1990. *Nature reserves—Island theory and conservation practice.* Washington, D.C.: Smithsonian Institution Press.

Trombulak, S., R. Noss, and J. Strittholt. 1996. Obstacles to implementing the Wildlands Project vision. *Wild Earth* 5(4): 84–89.

U.S. Fish and Wildlife Service. 1994. *An ecosystem approach to fish and wildlife management.* Washington, D.C.: U.S. Fish and Wildlife Service.

———. 1997. *Klamath/Central Pacific Coast ecoregion restoration strategy.* Klamath Falls, Oreg.: Klamath Basin Ecosystem Restoration Office.

Vance-Borland, K., R. Noss, J. Strittholt, P. Frost, C. Carroll, and R. Nawa. 1996. A biodiversity conservation plan for the Klamath/Siskiyou Region. *Wild Earth* 5(4): 52–59.

Walters, J. 1996. A reality check for TWP. *Wild Earth* 5(4): 16.

Chapter 4

Bailey, R. G. 1995. *Description of the ecoregions of the United States.* 2d ed. Washington, D.C.: USDA Forest Service, Misc. Publ. 1391.

Burley, F. W. 1988. Monitoring biological diversity for setting priorities in conservation. In *BioDiversity*, ed. E. O. Wilson and F. M. Peter, 227–230. Washington, D.C.: National Academy Press.

Butterfield, B. R., B. Csuti, and J. M. Scott. 1994. Modeling vertebrate distributions for gap analysis. In *Mapping the diversity of nature*, ed. R. I. Miller, 53–68. London: Chapman and Hall.

Caicco, S. L., J. M. Scott, B. Butterfield, and B. Csuti. 1995. A gap analysis of the management status of the vegetation of Idaho (U.S.A.). *Conservation Biology* 9: 498–511.

Camm, J. D., S. Polasky, A. Solow, and B. Csuti. 1996. A note on optimal algorithms for reserve site selection. *Biological Conservation* 78: 353–355.

Church, R. L., D. M. Stoms, and F. W. Davis. 1996. Reserve selection as a maximal covering location problem. *Biological Conservation* 76: 105–112.

Csuti, B. 1996. Mapping animal distribution areas for gap analysis. In *Gap analysis: A landscape approach to biodiversity planning*, ed. J. M. Scott, T. H. Tear, and F. W. Davis, 135–145. Bethesda, Md.: American Society of Photogrammetry and Remote Sensing.

Csuti, B., and A. R. Kiester. 1996. Hierarchical gap analysis for identifying priority areas for biodiversity. In *Gap analysis: A landscape approach to biodiversity planning*, ed. J. M. Scott, T. H. Tear, and F. W. Davis, 25–37.

Bethesda, Md.: American Society for Photogrammetry and Remote Sensing.

Csuti, B., S. Polasky, P. H. Williams, R. L. Pressey, J. D. Camm, M. Kershaw, A. R. Kiester, B. Downs, R. Hamilton, M. Huso, and K. Sahr. 1997. A comparison of reserve selection algorithms using data on terrestrial vertebrates in Oregon. *Biological Conservation* 80: 83–97.

Dasmann, R. F. 1972. Towards a system for classifying natural regions of the world and their representation by national parks and reserves. *Biological Conservation* 4: 247–255.

Davis, F. W., P. A. Stine, D. M. Stoms, M. I. Borchert, and A. D. Hollander. 1995. Gap analysis of the actual vegetation of California: 1. The southwestern region. *Madroño* 42: 40–78.

Davis, F. W., and D. M. Stoms. 1996. Sierran vegetation: A gap analysis. In *Assessments and scientific basis for management options, vol. II of Sierra Nevada Ecosystem Project, Final report to Congress*, 671–686. Davis: University of California, Centers for Water and Wildland Resources.

Davis, F. W., D. M. Stoms, R. L. Church, and W. J. Orkin. In press. *Balancing efficiency and suitability in siting representative systems of nature reserves.*

Davis, F. W., D. M. Stoms, R. L. Church, W. J. Orkin, and K. N. Johnson. 1996. Selecting biodiversity management areas. In *Assessments and scientific basis for management options, vol. II of Sierra Nevada Ecosystem Project, Final report to Congress*, 1503–1528. Davis: University of California, Centers for Water and Wildland Resources.

Driscoll, R. S., D. L. Merkel, D. L. Radloff, D. E. Snyder, and J. S. Hagihara. 1984. *An ecological land classification framework for the United States.* Washington, D.C.: USDA Forest Service, Misc. Publ. 1439.

Edwards, T. C., Jr., E. T. Deshler, D. Foster, and G. G. Moisen. 1996. Adequacy of wildlife habitat relation models for estimating spatial distributions of terrestrial vertebrates. *Conservation Biology* 10: 263–270.

Edwards, T. C., Jr., C. G. Homer, S. D. Bassett, A. Falconer, R. D. Ramsey, and D. W. Wright. 1995. *Utah gap analysis: An environmental information system.* Logan, Utah: Utah Cooperative Fish and Wildlife Research Unit, Utah State University, Technical Report 95-1.

Hickman, J. C. 1993. *The Jepson manual: Higher plants of California.* Berkeley: University of California Press.

Kendeigh, S. C., H. I. Baldwin, V. H. Cahalane, C. H. D. Clarke, C. Cottam, W. P. Cottam, I. McT. Cowan, P. Dansereau, J. H. Davis, Jr., F. W. Emerson, I. T. Haig, A. Hayden, C. L. Hayward, J. M. Linsdale, J. A. Macnab, and J. E. Potzger. 1950–51. Native santuaries in the United States and Canada, 1950–51. *The Living Wilderness* 15: 1–45.

Kiester, A. R., J. M. Scott, B. Csuti, R. F. Noss, B. Butterfield, K. Sahr, and D. White. 1996. Conservation prioritization using GAP data. *Conservation Biology* 10: 1332–1342.

Lawton, J. H., J. R. Prendergast, and B. C. Eversham. 1994. The numbers and spatial distributions of species: Analyses of British data. In *Systematics and conservation evaluation*, ed. P. L. Forey, C. J. Humphries, and R. I.

Vane-Wright. Systematics and Conservation Evaluation Special Volume 50, of the Systematics Association, 177–195. Oxford: Clarendon Press.

Loveland, T. R., and D. M. Shaw. 1996. Multi-resolution land characterization: Building collaborative partnerships. In *Gap analysis: A landscape approach to biodiversity planning*, ed. J. M. Scott, T. H. Tear, and F. W. Davis, 79–85. Bethesda, Md.: American Society for Photogrammetry and Remote Sensing.

Noss, R. F. 1987. From plant communities to landscapes in conservation inventories: A look at The Nature Conservancy (USA). *Biological Conservation* 41: 11–37.

_____. 1996. Protected areas: How much is enough? In *National parks and protected areas: Their role in environmental protection*, ed. R. G. Wright, 89–120. Cambridge, Mass.: Blackwell Science.

Noss, R. F., E. T. LaRoe III, and J. M. Scott. 1995. *Endangered ecosystems of the United States: A preliminary assessment of loss and degradation.* Washington, D.C.: USDI, National Biological Service, Biological Report No. 28.

Oregon Biodiversity Project. 1998. *Oregon's living landscape: Strategies and opportunities to conserve biodiversity.* Portland, Oregon: Defenders of Wildlife.

Orians, G. H. 1993. Endangered at what level? *Ecological Applications* 3: 206–208.

Pickett, S. T. A., and J. N. Thompson. 1978. Patch dynamics and the design of nature reserves. *Biological Conservation* 23–37.

Prendergast, J. R., R. M. Quinn, J. H. Lawton, B. C. Eversham, and D. W. Gibbons. 1993. Rare species, the coincidence of diversity hotspots and conservation strategies. *Nature* 365: 335–337.

Pressey, R. L. 1994. Ad hoc reservations: Forward or backward steps in developing representative reserve systems? *Conservation Biology* 8: 662–668.

Pressey, R. L., C. J. Humphries, C. R. Margules, R. I. Vane-Wright, and P. H. Williams. 1993. Beyond opportunism: Key principles for systematic reserve selection. *Trends in Ecology and Evolution* 8: 124–128.

Pressey, R. L., I. R. Johnson, and P. D. Wilson. 1994. Shades of irreplaceability: Towards a measure of the contribution of sites to a reservation goal. *Biodiversity and Conservation* 3: 242–262.

Pressey, R. L., H. P. Possingham, and C. R. Margules. 1996. Optimality in reserve selection algorithms: When does it matter and how much? *Biological Conservation* 76: 259–267.

Ride, W. L. D. 1975. Towards an integrated system: A study of selection and acquisition of national parks and nature reserves in Western Australia. In *Reports of the Australian Academy of Science*, no. 19, *A national system of ecological reserves in Australia*, ed. F. Fenner, 64–85. Canberra: Australian Academy of Science.

Scott, J. M., and B. Csuti. 1997a. Gap analysis for biodiversity survey and maintenance. In *Biodiversity II: Understanding and protecting our biologi-*

cal resources, ed. M. L. Reaka-Kudla, D. E. Wilson, and E. O. Wilson, 321–340. Washington, D.C.: Joseph Henry Press.

_____. 1997b. Noah worked two jobs. *Conservation Biology* 11: 1255–1257.

Scott, J. M., B. Csuti, J. D. Jacobi, and J. E. Estes. 1987. Species richness: A geographic approach to protecting future biological diversity. *BioScience* 37: 782–788.

Scott, J. M., B. Csuti, and K. A. Smith. 1990. Playing Noah while paying the Devil. *Bulletin of the Ecological Society of America* 71: 156–158.

Scott, J. M., F. Davis, B. Csuti, R. Noss, B. Butterfield, C. Groves, H. Anderson, S. Caicco, F. D'Erchia, T. C. Edwards, Jr., J. Ulliman, and R. G. Wright. 1993. Gap analysis: A geographic approach to protection of biological diversity. *Wildlife Monographs* 123: 1–41.

Scott, J. M., T. H. Tear, and F. W. Davis, eds. 1996. *Gap analysis: A landscape approach to biodiversity planning*. Bethesda, Md.: American Society for Photogrammetry and Remote Sensing.

Shaffer, M. L. 1981. Minimum population sizes for species conservation. *BioScience* 31: 131–134.

Specht, R. L., E. M. Roe, and V. H. Boughton. 1974. Conservation of major plant communities in Australia and Papua New Guinea. *Australian Journal of Botany*, supp. no. 7. Melbourne: Commonwealth Scientific and Industrial Research Organization.

Steinitz, C., M. Binford, P. Cote, T. Edwards, Jr., S. Ervin, R. T. T. Forman, C. Johnson, R. Kiester, D. Mouat, D. Olson, A. Shearer, R. Toth, and R. Wills. 1996. *Biodiversity and landscape planning: Alternative futures for the region of Camp Pendleton, California*. Privately published. Harvard University, Graduate School of Design.

Stine, P. A., F. W. Davis, B. Csuti, and J. M. Scott. 1996. Comparative utility of vegetation maps of different resolutions for conservation planning. In *Biodiversity in managed landscapes: Theory and practice*, ed. R. C. Szaro and D. W. Johnston, 210–220. New York: Oxford University Press.

Stokland, J. N. 1997. Representativeness and efficiency of bird and insect conservation in Norwegian boreal forest reserves. *Conservation Biology* 11: 101–111.

Stoms, D. M., F. W. Davis, K. L. Driese, K. M. Cassidy, and M. P. Murray. 1998. Gap analysis of the vegetation of the Intermountain Semi-Desert ecoregion. *Great Basin Naturalist* 58: 199–216.

Terborgh, J., and B. Winter. 1983. A method for siting parks and reserves with special reference to Colombia and Ecuador. *Biological Conservation* 27: 45–58.

Underhill, L. G. 1994. Optimal and suboptimal reserve selection algorithms. *Biological Conservation* 70: 85–87.

United Nations Educational, Scientific, and Cultural Organization (UNESCO). 1973. *International classification and mapping of vegetation*. Paris, France: UNESCO.

U.S. Department of Agriculture (USDA) Forest Service. 1996. *Status of the interior Columbia basin: Summary of scientific findings*. Portland, Ore.:

USDA Forest Service, Pacific Northwest Research Station; and USDI, Bureau of Land Management. General Technical Report PNW-GTR-385.

Williams, P., D. Gibbons, C. Margules, A. Rebelo, C. Humphries, and R. Pressey. 1996. A comparison of richness hotspots, rarity hotspots, and complementary areas for conserving diversity of British birds. *Conservation Biology* 10: 155–174.

Wright, R. G., J. G. MacCracken, and J. Hall. 1994. An ecological evaluation of proposed new conservation areas in Idaho: Evaluating proposed Idaho National Parks. *Conservation Biology* 8: 207–216.

Chapter 5

Agee, J. K. 1993. *Fire ecology of Pacific Northwest forests.* Washington, D.C.: Island Press.

Anderson, M. K., and M. J. Moratto. 1996. Native American land-use practices and ecological impacts. Ch. 9 in *Assessments and scientific basis for management options*, vol. II of *Sierra Nevada Ecosystem Project, Final report to Congress*, 187–206. Davis: University of California, Centers for Water and Wildland Resources.

Aplet, G. H., N. Johnson, J. T. Olson, and V. A. Sample. 1993. Prospects for a sustainable future. In *Defining sustainable forestry*, ed. G. H. Aplet, N. Johnson, J. T. Olson, and V. A. Sample, 309–314. Washington, D.C.: Island Press.

Baker, W. L. 1992. The landscape ecology of large disturbances in the design and management of nature reserves. *Landscape Ecology* 7: 181–194.

Betancourt, J. L., and T. R. Van Devender. 1981. Holocene vegetation in Chaco Canyon, New Mexico. *Science* 214: 656–658.

Botkin, D. B. 1990. *Discordant harmonies: A new ecology for the twenty-first century.* New York: Oxford University Press.

Camp, A. 1995. Predicting late-successional fire refugia from physiography and topography. Ph.D. diss., University of Washington.

Caraher, D. L., J. Henshaw, F. Hall, W. H. Knapp, B. P. McCammon, J. Nesbitt, R. J. Pedersen, I. Regenovitch, and C. Tietz. 1992. *Restoring ecosystems in the Blue Mountains: A report to the regional forester and the forest supervisors of the Blue Mountain forests.* Portland, Ore.: USDA Forest Service, Pacific Northwest Region.

Chapel, M., M. Flores, D. G. Fuller, P. Manley, and C. Millar. 1995. Conserving biological diversity: The coarse filter/fine filter approach. Appendix I-B in *Sustaining ecosystems: A conceptual framework*, ed. P. Manley, G. E. Brogan, C. Cook, M. E. Flores, D. G. Fullmer, S. Husari, T. M. Jimerson, L. M. Lux, M. E. McCain, J. A. Rose, G. Schmitt, J. C. Schuyler, and M. J. Skinner. San Francisco: USDA Forest Service, Pacific Southwest Region, R5-EM-TP-001.

Christensen, N. L. 1988. Succession and natural disturbance: Paradigms, problems, and preservation of natural ecosystems. In *Ecosystem management for parks and wilderness*, ed. J. K. Agee and D. R. Johnson, 62–86. Seattle: University of Washington Press.

Clements, F. E. 1916. *Plant succession: An analysis of the development of vegetation.* Washington, D.C.: Carnegie Institution of Washington.

Comer, P. J. 1997. Letter to the editor. *Conservation Biology* 11: 301–303.

Cronon, W. 1983. *Changes in the land: Indians, colonists, and the ecology of New England.* New York: Hill and Wang.

DellaSala, D. A., J. R. Strittholt, R. F. Noss, and D. M. Olson. 1996. A critical role for core reserves in managing inland northwest landscapes for natural resources and biodiversity. *Wildlife Society Bulletin* 24: 209–221.

FEMAT (Forest Ecosystem Management Assessment Team). 1993. *Forest ecosystem management: An ecological, economic, and social assessment.* Report of the Forest Ecosystem Management Assessment Team. Washington, D.C.: USDA Forest Service, USDC National Marine Fisheries Service, USDI Bureau of Land Management, USDI Fish and Wildlife Service, USDI National Park Service, and the Environmental Protection Agency.

Foreman, D. 1995–96. Wilderness: From scenery to nature. *Wild Earth* 1995/96: 8–16.

Franklin, J. F. 1993. The fundamentals of ecosystem management with applications in the Pacific Northwest. In *Defining sustainable forestry*, ed. G. H. Aplet, N. Johnson, J. T. Olson, and V. A. Sample, 127–144. Washington, D.C.: Island Press.

Frissell, C. A., and D. Bayles. 1996. Ecosystem management and the conservation of aquatic biodiversity and ecological integrity. *Water Resources Bulletin* 32: 229–240.

Hansen, A. J., T. A. Spies, F. J. Swanson, and J. L. Ohmann. 1991. Conserving biodiversity in managed forests. *BioScience* 41: 382–392.

Harting, A., and D. Glick. 1994. *Sustaining Greater Yellowstone, a blueprint for the future.* Bozeman, Mont.: Greater Yellowstone Coalition.

Haufler, J. B. 1994. An ecological framework for planning for forest health. *Journal of Sustainable Forestry* 2: 307–316.

Hunter, M. L. 1991. Coping with ignorance: The coarse-filter strategy for maintaining biodiversity. In *Balancing on the brink of extinction the Endangered Species Act and lessons for the future*, ed. K. A. Kohm, 266–281. Washington, D.C.: Island Press.

_____. 1996. Benchmarks for managing ecosystems: Are human activities natural? *Conservation Biology* 10: 695–697.

Jenny, H. 1941. *Factors of soil formation.* New York: McGraw-Hill.

_____. 1980. *Soil genesis with ecological perspectives.* New York: Springer-Verlag.

Johnson, C. G., R. R. Clausnitzer, P. J. Mehringer, and C. D. Oliver. 1994. *Biotic and abiotic processes of eastside ecosystems: The effects of management on plant and community ecology, and on stand and landscape vegetation dynamics.* USDA Forest Service, General Technical Report PNW-GTR-322.

Keeton, W. S., and G. H. Aplet. 1997. *Ecosystem management in the Interior Columbia River Basin.* Seattle: The Wilderness Society.

Landres, P. B., P. S. White, G. Aplet, and A. Zimmerman. 1998. Naturalness and natural variability: Definitions, concepts, and strategies for wilderness management. In *Proceedings of the Second Eastern Wilderness Conference*, ed. D. L. Kulhavy. Nacogdoches, Tex.: Center for Applied Studies, School of Forestry, Stephen F. Austin University.

Leopold, A. 1949. *A Sand County almanac.* New York: Oxford University Press.

Manley, P., G. E. Brogan, C. Cook, M. E. Flores, D. G. Fullmer, S. Husari, T. M. Jimerson, L. M. Lux, M. E. McCain, J. A. Rose, G. Schmitt, J. C. Schuyler, and M. J. Skinner. 1995. *Sustaining ecosystems: A conceptual framework.* San Francisco: USDA Forest Service, Pacific Southwest Region, R5-EM-TP-001.

Millar, C. I. 1997. Commentary on historical variation and desired condition as tools for terrestrial landscape analysis. In *What is watershed stability? Proceedings of the 6th Biennial Watershed Management Conference*, ed. S. Sommarstrom. (October): 23–25, 1996. Lake Tahoe, Cal.: University of California.

Mooney, H. A., T. M. Bonnicksen, N. L. Christensen, J. E. Lotan, and W. A. Reiners, eds. 1981. *Proceedings of the fire regimes and ecosystem properties conference,* December 11–15, 1978, Honolulu, Hawaii.

Morgan, P., G. H. Aplet, J. B. Haufler, H. C. Humphries, M. M. Moore, and W. D. Wilson. 1994. Historical range of variability: A useful tool for evaluating ecosystem change. *Journal of Sustainable Forestry* 2: 87–111.

Morrison, P. H., and F. J. Swanson. 1990. *Fire history and pattern in a Cascade Range landscape.* USDA Forest Service, General Technical Report PNW-GTR-254.

Nature Conservancy, The. 1982. *Natural Heritage Program operations manual.* Arlington, Va.: The Nature Conservancy.

Noss, R. F., and A. Y. Cooperrider. 1994. *Saving nature's legacy.* Washington, D.C.: Island Press.

Office of Technology Assessment (OTA). 1993. *Harmful non-indigenous species in the United States.* Washington, D.C.: Government Printing Office, OTA-F-565.

Parsons, D. J. 1981. The role of fire management in maintaining natural ecosystems. *Proceedings of the fire regimes and ecosystem properties conference*, December 11–15, 1978, Honolulu, Hawaii. Tech. coord., H. A. Mooney, T. M. Bonnicksen, N. L. Christensen, J. E. Lotan, and W. A. Reiners, 469–488. USDA Forest Service, General Technical Report WO-26.

Pickett, S. T. A., V. T. Parker, and P. L. Fiedler. 1992. The new paradigm in ecology: Implications for conservation biology above the species level. In *Conservation biology: The theory and practice of nature conservation, preservation, and management*, ed. P. L. Fiedler and S. K. Jain, 65–88. New York: Chapman and Hall.

Pickett, S. T. A., and J. N. Thompson. 1978. Patch dynamics and the design of nature reserves. *Biological Conservation* 13: 27–37.

Pickett, S. T. A., and P. S. White, eds. 1985. *The ecology of natural disturbance and patch dynamics.* Orlando, Fla.: Academic Press.

Rhodes, J. J., D. A. McCullough, and F. A. Espinosa, Jr. 1994. *A coarse screening process for evaluation of the effects of land management activities on salmon spawning and rearing habitat in ESA consultations.* Portland, Ore.: Columbia River Intertribal Fish Commission, Technical Report 94-4.

Robbins, C. S., D. K. Dawson, and B. A. Dowell. 1989. Habitat area requirements of breeding forest birds of the Middle Atlantic states. *Wildlife Monographs* 103: 1–34.

Schrader-Frechette, K. S., and E. D. McCoy. 1995. Natural landscapes, natural communities, and natural ecosystems. *Forest and Conservation History* 39: 138–142.

Shaffer, M. 1981. Minimum population sizes for species conservation. *BioScience* 31: 131–134.

Shelford, V. E., ed. 1926. *Naturalist's guide to the Americas.* Baltimore, Md.: Williams and Wilkins.

Shugart, H. H., and D. C. West. 1981. Long-term dynamics of forest ecosystems. *American Scientist* 69: 647–652.

Simberloff, D., J. A. Farr, J. Cox, and D. W. Mehlman. 1992. Movement corridors: Conservation bargains or poor investments. *Conservation Biology* 6: 493–504.

Soulé, M. E., and D. Simberloff. 1986. What do genetics and ecology tell us about the design of nature reserves? *Biological Conservation* 35: 19–40.

Sullivan, A. L., and M. L. Shaffer. 1975. Biogeography of the megazoo. *Science* 189: 13–17.

Swanson, F. J., J. A. Jones, D. O. Wallin, and J. H. Cissel. 1994. Natural variability—Implications for ecosystem management. In *Ecosystem management: Principles and applications,* vol. II of the *Eastside Forest Ecosystem Health Assessment* (R. Everett, Team Leader), 80–94. USDA Forest Service, General Technical Report PNW-GTR-318.

Thomas, J. W., E. D. Forsman, J. B. Lint, E. C. Meslow, B. R. Noon, and J. Verner. 1990. *A conservation strategy for the northern spotted owl.* Washington, D.C.: USDA Forest Service, USDI Bureau of Land Management, USDI National Park Service.

Thomas, W. L., Jr., ed. 1956. *Man's role in changing the face of the earth.* Chicago: University of Chicago Press.

Vitousek, P. M. 1986. Biological invasions and ecosystem properties: Can species make a difference? In *Ecology of biological invasions of North America and Hawaii,* ed. H. A. Mooney and J. A. Drake, 163–176. New York: Springer-Verlag.

Watt, A. S. 1947. Pattern and process in the plant community. *Journal of Ecology* 35: 12–22.

Wegner, D. 1996. Floods in the Grand Canyon: The first step towards understanding critical ecosystem process below Glen Canyon Dam, Ariz. Flagstaff: Glen Canyon Environmental Studies.

White, P. S., and S. T. A. Pickett. 1985. Natural disturbance and patch dynamics: An introduction. In *The ecology of natural disturbance and patch dynamics,* ed. S. T. A. Pickett and P. S. White. Orlando, Fla.: Academic Press.

Whitney, G. G. 1994. *From coastal wilderness to fruited plain: A history of environmental change in temperate North America, 1500 to the present.* New York: Cambridge University Press.

Woolfenden, W. B. 1996. Quaternary vegetation history. Ch. 5 in *Assessments and scientific basis for management options,* vol. II of *Sierra Nevada Ecosystem Project, Final report to Congress.* Davis: University of California, Centers for Water and Wildland Resources.

Chapter 6

Agee, J. K., and R. L. Edmonds. 1992. Forest protection issues for the northern spotted owl. Appendix E of *Final draft recovery plan for the northern spotted owl.* Washington, D.C.: USDI.

Agee, J. K., and M. H. Huff. 1985. Structure and process goals for vegetation in wilderness areas. *National Wilderness Research Conference,* 17–25. Ogden, Utah: USDA Forest Service, Intermountain Research Station, General Technical Report INT-212.

Agee, J. K., and D. R. Johnson. 1988. *Ecosystem management for parks and wilderness: Workshop synthesis.* Institute of Forest Resources, Cont. 62, 1–19. Seattle: University of Washington.

Averill, R. D., L. Larson, J. Saveland, P. Wargo, and J. Williams. 1997. *Disturbance processes and ecosystem management.* USDA Forest Service, white paper.

Brunson, M. W. In press. Social dimensions of boundaries: Balancing cooperation and self-interest. In *Stewardship across boundaries,* ed. R. L. Knight, and P. B. Landres. Washington, D.C.: Island Press.

Camp, A. 1995. Predicting late-successional fire refugia from physiography and topography. Ph.D. diss., University of Washington.

Camp, A. E., P. F. Hessburg, and R. L. Everett. 1996. Dynamically incorporating late-successional forest in sustainable landscapes. In *The use of fire in forest restoration,* ed. C. Hardy and S. Arno. Ogden, Utah: USDA Forest Service, Intermountain Research Station, General Technical Report INT-341.

Christensen, N. L. 1988. Succession, and natural disturbance paradigms, problems and preservation of natural ecosystems. In *Ecosystem management for parks and wilderness,* ed. J. K. Agee and D. R. Johnson, 62–86. Seattle: University of Washington.

Covington, W., R. L. Everett, R. Steele, L. L. Irwin, T. Daer, and A. Auclair. 1994. Historical and anticipated changes in forest ecosystems of the inland west of the United States. *Journal of Sustainable Forestry* 2(1/2): 13–63.

DellaSala, D. A., J. R. Strittholt, R. F. Noss, and D. M. Olson. 1996. A critical role for core reserves in managing inland northwest landscapes for natural resources and biodiversity. *Wildlife Society Bulletin* 24(2): 209–221.

Diaz, N., and D. Apostol. 1993. *Forest landscape analysis and design: A process for developing and implementing land management objectives for landscape patterns.* Portland, Ore.: USDA Forest Service, Pacific Northwest Region, R6 ECO-TP-043-92.

Dombeck, M. P. 1996. Thinking like a mountain: BLM's approach to ecosystem management. *Ecological Applications* 6: 699–702.

Ehrenfeld, D. W. 1991. The management of diversity: A conservation paradox. In *Ecology, economics, ethics: The broken circle*, ed. F. H. Bormann and S. H. Kellert, 26–39. New Haven: Yale University Press.

Everett, R., P. Hessburg, J. Lehmkuhl, and P. Bourgeron. 1994. Old forests in dynamic landscapes—Dry site forests of eastern Oregon and Washington. *Journal of Forestry* 92: 22–25.

Everett, R. L., P. F. Hessburg, and T. R. Lillybridge. 1994. Emphasis areas as an alternative to buffer zones and reserved areas in the conservation of biodiversity and ecosystem processes. *Journal of Sustainable Forestry* 2(3/4): 283–292.

Everett, R. L., and J. F. Lehmkuhl. 1996. An emphasis-use approach to conserving biodiversity. *Wildlife Bulletin* 24(2): 192–199.

Everett, R. L., R. Schellhaas, T. Anderson, J. Lehmkuhl, and A. Camp. 1996. Restoration of ecosystem integrity and land use allocations objective in altered watersheds. In *Watershed restoration management*, ed. J. J. McDonald, J. B. Stribling, L. R. Neville, and D. J. Leopold, 271–280. Herndon, Va.: Watershed Restoration Management, American Water Resources Association.

Everett, R., R. Schellhaas, D. Spurbeck, P. Ohlson, D. Keenum, and T. Anderson. 1997. Structure of northern spotted owl nest stands and their historical conditions on the eastern slope of the Pacific Northwest Cascades, USA. *Forest Ecology and Management* 94: 1–14.

Everett, R. L., J. Townsley, and D. M. Baumgartner. In press. Inherent disturbance regimes: A reference for evaluating long-term maintenance of ecosystems. *Journal of Sustainable Forestry*.

Fellers, G. M., and V. Norris. 1991. Rare plant monitoring and management at Point Reyes. *A Resource Management Bulletin* 11: 20–21. USDI, National Park Service.

Franklin, J. F., and R. T. T. Forman. 1987. Creating landscape patterns by forest cutting: Ecological consequences and principles. *Landscape Ecology* 1: 5–18.

Fritts, H. C., and T. W. Swetnam. 1989. Dendroecology: A tool for evaluating variations in past and present forest environments. *Advanced Ecological Research* 19: 11–18.

Grier, C. 1975. Wildfire effects on nutrient distribution and leaching in a coniferous ecosystem. *Canadian Journal of Forest Research* 5: 599–607.

Grier, C. C., K. M. Lee, N. M. Nadkarni, G. O. Klock, and P. J. Edgerton. 1989. *Productivity of forests of the United States and its relation to soil and site factors and management practices: A review*. Portland, Ore.: USDA Forest Service, Pacific Northwest Research Station. General Technical Report PNW-GTR-222: 51.

Hansen, A. J., T. A. Spies, F. J. Swanson, and J. L. Ohmann. 1991. Conserving biodiversity in managed forests: Lessons from natural forests. *BioScience* 41: 382–391.

Harris, L. D. 1984. *The fragmented forest.* Chicago: University of Chicago Press.

Harvey, A., M. Larsen, and M. Jurgensen. 1981. *Rate of woody residue incorporation in Northern Rocky Mountain forest soils.* Ogden, Utah: USDA Forest Service, Intermountain Research Station, Research Paper INT-282.

Haufler, J. B., C. A. Mehl, and G. J. Roloff. 1996. Using a coarse-filter approach with species assessment for ecosystem management. *Wildlife Society Bulletin* 24(2): 200–208.

Haynes, R. W., R. T. Graham, and T. M. Quigley. 1996. *A framework for ecosystem management in the interior Columbia Basin.* Portland, Ore.: USDA Forest Service, Pacific Northwest Research Station, General Technical Report PNW-374.

Hessburg, P. F., R. G. Mitchell, and G. M. Filip. 1994. *Historical and current roles of insects and pathogens in eastern Oregon and Washington forest landscapes.* Portland, Ore.: USDA Forest Service, Pacific Northwest Research Station, General Technical Report PNW-327.

Hunter, M. L., Jr. 1987. Managing forest for spatial heterogeneity to maintain biological diversity. In *Transactions of the North American Wildfire and Natural Resources Conference* 52: 60–69.

_____. 1993. Natural fire regimes as spatial models for managing boreal forests. *Biological Conservation* 65: 115–120.

Jansen, D. H. 1986. The eternal external threat. Ch. 13 in *Conservation Biology: The science of scarcity and diversity,* ed. M. E. Soulé, 286–303. Sunderland, Mass.: Sinauer Associates.

Keiter, R. 1988. Natural ecosystem management in park and wilderness areas: Looking at the law. In *Ecosystem management for parks and wilderness—Workshop synthesis,* ed. J. K. Agee and D. R. Johnson. Institute of Forest Resources, cont. No. 62. Seattle: University of Washington.

Landres, P. B., R. L. Knight, S. T. A. Pickett, and M. L. Cadenasso. 1998a. Ecological effects of administrative boundaries. In *Stewardship across boundaries,* ed. R. L. Knight and P. B. Landres. Washington, D.C.: Island Press.

Landres, P. B., S. Marsh, L. Merigliano, D. Ritter, and A. Norman. 1998b. Boundary effects on National Forest wildernesses and other natural areas. In *Stewardship across boundaries,* ed. R. L. Knight and P. B. Landres. Washington, D.C.: Island Press.

Lehmkuhl, J. F., M. G. Raphael, R. S. Holthausen, J. R. Hickenbottom, R. H. Naney, and J. S. Shelly. 1997. Historical and current status of terrestrial species and the effects of proposed alternatives, 537–730. In *Evaluation of EIS alternatives by the Science Integration Team,* volume II, ed. T. M. Quigley, K. M. Lee, and S. J. Arbelbide. USDA Forest Service General Technical Report PNW-GTR-406. Portland, Ore.: Pacific Northwest Research Station.

Leopold, A. S., S. A. Cain, C. H. Cottam, I. N. Gabrielson, and T. L. Kimball. 1963. Wildlife management in the national parks. *American Forester* 69(4): 32–35, 61–63.

Lundquist, J. E. 1995. Disturbance profile-measure small-scale disturbance patterns in ponderosa pine stands. *Forest Ecology and Management* 74: 49–59.

Morgan, P., G. H. Aplet, J. B. Haufler, H. C. Humphries, M. M. Moore, and W. D. Wilson. 1994. Historical range of variability: A useful tool for evaluating ecosystem change. *Journal of Sustainable Forestry* 2: 87–112.

Noss, R. F. 1983. A regional landscape approach to maintain diversity. *BioScience* 33: 700–706.

_____. 1987. From plant communities to landscapes in conservation inventories: A look at The Nature Conservancy. *Biological Conservation* 41: 11–37.

_____. 1993. A conservation plan for the Oregon coast range: Some preliminary suggestions. *Natural Areas Journal* 13: 276–290.

Noss, R. F., and A. Y. Cooperrider. 1994. *Saving nature's legacy.* Washington, D.C.: Island Press.

Odum, E. P. 1969. The strategy of ecosystem development. *Science* 16: 262–270.

Overbay, J. C. 1992. Ecosystem management. *Taking an ecological approach to management. Proceedings of the national workshop*, April 27–30, 1992, Salt Lake City, Utah, 3–15. Washington, D.C.: U.S. Forest Service, WO-WSA-3.

Parsons, D. J., D. M. Graber, J. K. Agee, and J. W. van Wagtendonk. 1986. Natural fire management in national parks. *Environmental Management* 10: 21–24.

Pickett, S. T. A., and J. N. Thompson. 1978. Patch dynamics and the design of nature reserves. *Biological Conservation* 13: 27–37.

Quigley, T. M., and H. Bigler Cole. 1997. *Highlighted scientific findings of the Interior Columbia Basin Ecosystem Management Project.* Portland, Ore.: USDA Forest Service, Pacific Northwest Station; USDI, Bureau of Land Management, General Technical Report PNW-GTR-404.

Quigley, T. M., R. W. Haynes, and R. T. Graham. 1996. *Integrated scientific assessment for ecosystem management in the interior Columbia Basin.* Portland, Ore.: USDA Forest Service, Pacific Northwest Research Station, General Technical Report PNW-382.

Samson, F. B. 1992. Conserving biological diversity in sustainable ecological systems. *Transactions of the Fifty-Seventh North America Wildlife and Natural Resource Conference* 57: 308–319.

Schonewald-Cox, C. M., and J. W. Bayless. 1986. The boundary model: A geographical analysis of design and conservation of nature reserves. *Biological Conservation* 38: 305–322.

Smith, C. L., B. S. Steel, P. C. List, and S. Cordray. 1995. Making forest policy: Integrating GIS with social processes. *Journal of Forestry* 31–36.

Soulé, M. E. 1986. The effects of fragmentation. In *Conservation Biology*, 2d ed., ed. M. E. Soulé, 233–236. Sunderland, Mass.: Sinauer Associates.

Soulé, M. E., and B. A. Wilcox. 1980. *Conservation biology.* Sunderland, Mass.: Sinauer Associates.

Spies, T. A., J. F. Franklin, and M. Klopsch. 1990. Canopy gaps in Douglas-fir forest of the Cascade Mountains. *Canadian Journal of Forest Research* 20: 649–658.

Sprugel, D. G. 1991. Disturbance, equilibrium, and environmental variability: What is "natural" vegetation in a changing environment? *Biological Conservation* 58: 1–18.

Stoszek, K. J. 1990. Managing change through adaptive forestry. *Focus on Renewable Natural Resources* 15: 1–2. University of Idaho, Department of Forestry.

Swanson, F. J., and J. F. Franklin. 1992. New forestry principles from ecosystem management analysis of Pacific Northwest forests. *Ecological Applications* 2: 262–274.

Swanson, F. J., J. A. Jones, D. O. Wallin, and J. H. Cissel. 1994. Natural variability-implications for ecosystem management. In *Ecosystem management: Principles and applications*. Portland, Ore.: USDA Forest Service, Pacific Northwest Research Station, General Technical Report PNW-318.

Thomas, J. W., ed. 1979. Wildlife habitats in managed forests: The Blue Mountains of Oregon and Washington. *Agriculture Handbook*, no. 533. Washington, D.C.: USDA Forest Service.

Turner, M. G., W. H. Romme, R. H. Gardner, R. V. O'Neill, and T. K. Kratz. 1993. A revised concept of landscape equilibrium: Disturbance and stability on scaled landscapes. *Landscape Ecology* 8: 213–227.

Uebelacker, M. L. 1986. *Geographic explorations in the southern Cascades or eastern Washington: Changing land, people and resources.* Ph.D. diss., University of Oregon.

Urban, D. L., R. V. O'Neil, and H. H. Shugart, Jr. 1987. Landscape ecology: A hierarchical perspective can help scientists understand spatial patterns. *BioScience* 37: 119–127.

U. S. Department of Agriculture (USDA). 1993. *Forest ecosystem management: An ecological, economic and social assessment.* A report of the Forest Ecosystem Management Assessment Team, USDA, USDC, USDI, and EPA, Portland, Ore.

U. S. Department of Agriculture (USDA) and U. S. Department of Interior (USDI). 1994. *Environmental assessment for the implementation of interim strategies for managing anadromous fish-producing watersheds in eastern Oregon and Washington, Idaho, and portions of California.* Washington, D.C.: USDA Forest Service and USDI, Bureau of Land Management.

Walker, B. H. 1992. Biodiversity and ecological redundancy. *Conservation Biology* 6(1): 18–23.

Wargo, P. M. 1995. Disturbance in forest ecosystems caused by pathogen and insects. *Proceedings: Forest Health through Silviculture*, 1995 National Silviculture Workshop, Mescalero, N. Mex., 20–25. Fort Collins, Colo.: USDA Forest Service, Rocky Mountain Forest and Range Experiment Station, General Technical Report RM-267.

White, P. S. 1987. Natural disturbance, patch dynamics, and landscape pattern in natural areas. *Natural Areas Journal* 7(1): 14–22.

Wiens, J. A., C. S. Crawford, and J. R. Gosz. 1985. Boundary dynamics: A conceptual framework for studying landscape ecosystems. *Oikos* 45: 421–427.

Chapter 7

Camp, A., C. Oliver, P. Hessburg, and R. Everett. 1997. Predicting late successional fire refugia predating European settlement in the Wenatchee Mountains. *Forest Ecology Management* 95: 63–77.

Daubenmire, R. F. 1952. Forest vegetation of northern Idaho and adjacent Washington, and its bearing on concepts of vegetation classification. *Ecological Monographs* 22: 301–330.

_____. 1968. *Plant communities: A textbook of plant synecology.* New York: Harper and Row.

Ecomap. 1993. *National hierarchical framework of ecological units.* Washington, D.C.: USDA Forest Service.

Haufler, J. B. 1994. An ecological framework for forest planning for forest health. *Journal of Sustainable Forestry* 2: 307–316.

_____. 1995. Forest industry partnerships for ecosystem management. *Transactions of the North American Wildlife and Natural Resources Conference* 60: 422–432.

Haufler, J. B., T. Crow, and D. Wilcove. In press. Scale considerations for ecosystem management. *Proceedings of Ecological Stewardship Workshop.*

Haufler, J. B., C. A. Mehl, and G. J. Roloff. 1996. Using a coarse-filter approach with a species assessment for ecosystem management. *Wildlife Society Bulletin* 24: 200–208.

Kaufmann, M. R., R. T. Graham, D. A. Boyce, Jr., W. H. Moir, L. Perry, R. T. Reynolds, R. Bassett, P. Mehlhop, C. B. Edminster, W. M. Block, and P. S. Corn. 1994. *An ecological basis for ecosystem management.* USDA Forest Service, General Technical Report RM-246.

McNab, W. H., and P. E. Avers, comps. 1994. *Ecological subregions of the United States: Section descriptions.* USDA Forest Service, Admin. Publ. WO-WSA-5.

Morrison, M. L., B. G. Marcot, and R. W. Mannan. 1992. *Wildlife-habitat relationships: Concepts and applications.* Madison: University of Wisconsin Press.

O'Hara, K. L., P. A. Latham, P. Hessburg, and B. G. Smith. 1996. A structural classification for inland northwest forest vegetation. *Western Journal of Applied Forestry* 11(3): 97–102.

Roloff, G. J., and J. B. Haufler. 1997. Establishing population viability planning objectives based on habitat potentials. *Wildlife Society Bulletin* 25: 895–904.

Steele, R., R. D. Pfister, R. A. Ryker, and J. A. Kittams. 1981. *Forest habitat types of central Idaho.* USDA Forest Service, General Technical Report INT-114.

Chapter 8

Boling, K. C., D. Murphy, M. Goodwin, and M. D. Sullivan. 1996. Landscape management in Idaho: Meeting the challenge through organizational and technological innovation. *Journal of Forestry.*

Haufler, J. B., C. A. Mehl, and G. J. Roloff. 1996. Using a coarse-filter approach with species assessment for ecosystem management. *Wildlife Society Bulletin* 24(2): 200–208.

Morgan, P., G. H. Aplet, J. B. Haufler, H. C. Humphries, M. M. Moore, and W. D. Wilson. 1994. Historical range of variability: A useful tool for evaluating ecosystem change. *Journal of Sustainable Forestry* 2: 87–112.

O'Hara, K. L., P. A. Latham, P. Hessburg, and B. G. Smith. 1996. A structural classification for inland northwest forest vegetation. *Western Journal of Applied Forestry* 11(3): 97–102.

Thomas, J. W., ed. 1979. Wildlife habitats in managed forests: The Blue Mountains of Oregon and Washington. *Agriculture Handbook*, no. 553. Washington, D.C.: USDA Forest Service.

Wykoff, W. R., N. L. Crookston, and A. R Stage. 1982. *User's guide to the stand prognosis model.* Ogden, Utah: USDA Forest Service, Intermountain Forest and Range Experiment Station, General Technical Report INT-133.

Chapter 9

Anonymous. 1986. *North American waterfowl management plan.* Washington, D.C. and Ottawa, Ontario: USDI and Canadian Department of Environment.

_____. 1994a. *Canadian Landbird Conservation Strategy: Draft for discussion.* Hull, Quebec: Environment Canada, Migratory Birds Conservation Division, Canadian Wildlife Service; Ottawa, Ontario: Canadian Nature Federation.

_____. 1994b. *1994 update to the North American Waterfowl Management Plan.* Ottawa, Ontario: Canadian Department of Environment; Washington, D.C.: USDI; and Mexico City, Mexico: Mexican Department of the Environment.

_____. 1996. *Framework for landbird conservation in Canada.* Hull, Quebec: Canadian Landbird Conservation Working Group.

Batt, B. D. J., M. G. Anderson, C. D. Anderson, and F. D. Caswell. 1989. The use of prairie potholes by North American ducks. In *Northern prairie wetlands*, ed. A. G. van der Valk, 204–227. Ames: Iowa State University Press.

Calverly, B. K., and T. Sankowski. 1995. *Effectiveness of tractor-mounted flushing devices in reducing accidental mortality of upland nesting ducks in Central-Alberta hayfields.* Edmonton: Alberta NAWMP Centre, NWAMP-019.

Cowardin, L. M., D. H. Johnson, T. L. Shaffer, and D. W. Sparling. 1988. *Application of a simulation model to decisions in mallard management.* U.S. Fish and Wildlife Service, Fish and Wildlife Technical Report 17.

Dale, B. 1994. *Summary of the North American Waterfowl Management Plan Assessment for non-game species in the three Prairie Provinces in 1993.* Edmonton, Alberta: Environment Canada, Canadian Wildlife Service, Prairie and Northern Region, unpublished report.

_____. 1995. *Rotational grazing and songbirds.* Edmonton, Alberta: Environment Canada, Canadian Wildlife Service, Prairie and Northern Region, unpublished report.

de Sobrino, C., and T. Arnold. 1995. *Progress report: Manitoba songbird project.* Washington, D.C.: Institute for Wetland and Waterfowl Research; and Oak Hammond, Manitoba: Ducks Unlimited Canada, unpublished progress report.

Dohl, S., J. Horton, and R. E. Jones. 1995. *1994 non-waterfowl evaluation of Manitoba's North American Waterfowl Management Plan.* Winnipeg: Manitoba Department of Natural Resources, unpublished report.

Duebbert, H. F., and H. A. Kantrud. 1987. Use of no-till winter wheat by nesting ducks in North Dakota. *Journal of Soil and Water Conservation* 42: 50–53.

Fisher, C. C., and W. Roberts. 1994. *Herpetofaunal investigations on North American Waterfowl Management Plan properties in Alberta—1994.* 1. Aspen parkland grassland restoration; 2. Medicine Wheel Landscape. Edmonton: Alberta NAWMP Centre, NAWMP-010.

Greenwood, R. J., A. B. Sargeant, D. H. Johnson, L. M. Cowardin, and T. L. Shaffer. 1995. Factors associated with duck nesting success in the Prairie Pothole Region of Canada. *Wildlife Monographs* 128: 1–57.

Grumbine, R. E. 1994. What is ecosystem management? *Conservation Biology* 8: 27–38.

Hartley, M. J. 1994. Passerine abundance and productivity indices in grasslands managed for waterfowl nesting cover. *Transactions of the Fifty-Ninth North American Wildlife and Natural Resources Conference* 57: 322–327.

Jacobsen, E. T., D. B. Wark, R. G. Arnott, R. J. Haas, and D. A. Tober. 1994. Sculptured seeding: An ecological approach to revegetation. *Restoration and Management Notes* 12(1): 46–50.

Johnson, D. H., J. D. Nichols, and M. D. Schwartz. 1992. Population dynamics of breeding waterfowl. In *Ecology and management of breeding waterfowl,* ed. B. D. J. Batt, A. D. Afton, M. G. Anderson, C. D. Ankney, D. H. Johnson, J. A. Kadlec, and G. L. Krapu, 446–485. Minneapolis: University of Minnesota Press.

Klett, A. T., T. L. Shaffer, and D. H. Johnson. 1988. Duck nest success in the Prairie Pothole Region. *Journal of Wildlife Management* 52: 431–440.

Prescott, D. R. C., A. J. Murphy, and E. Ewaschuk. 1995. *An avian community approach to determining biodiversity values of NAWMP habitats in the aspen parkland of Alberta.* Edmonton: Alberta NAWMP Centre, NAWMP-012.

Sargeant, A. B., R. J. Greenwood, M. A. Sovada, and T. L. Shaffer. 1993. *Distribution and abundance of predators that affect duck production— Prairie Pothole Region.* U.S. Fish and Wildlife Service, Resource Publ. 194.

Sargeant, A. B., and D. G. Raveling. 1992. Mortality during the breeding season. In *Ecology and management of breeding waterfowl,* ed. B. D. J. Batt, A. D. Afton, M. G. Anderson, C. D. Ankney, D. H. Johnson, J. A.

Kadlec, and G. L. Krapu, 396–422. Minneapolis: University of Minnesota Press.

Skinner, D. L., S. Johnston, and D. A. Westworth. 1995. *Abundance and distribution of mammals in planted nesting cover in the aspen parkland of Alberta: North American Waterfowl Management Plan.* Edmonton: D. A. Westworth and Associates Ltd. and Alberta NAWMP Centre, NAWMP-015.

Sovada, M. A., A. B. Sargeant, and J. W. Grier. 1995. Differential effects of coyotes and red foxes on duck nest success. *Journal of Wildlife Management* 59: 1–9.

Stewart, R. E., and H. A. Kantrud. 1971. *Classification of natural ponds and lakes in the glaciated prairie region.* U.S. Fish and Wildlife Service, Resource Publ. 92.

Trottier, G. C. 1992. *Conservation of Canadian prairie grasslands: A landowner's guide.* Edmonton, Alberta: Environment Canada, Canadian Wildlife Service, Prairie and Northern Region.

Wark, D. B., W. R. Poole, R. G. Arnott, L. R. Moats, and L. Wetter. 1995. *Revegetating with native grasses.* Oak Hammock, Manitoba: Ducks Unlimited Canada.

Chapter 10

Angermeier, P. L., and J. R. Karr. 1994. Biotic integrity versus biological diversity as policy directives: Protecting biotic resources. *BioScience* 44: 690–697.

Barbour, M. G., and W. D. Billings. 1993. *North American vegetation.* New York: Cambridge University Press.

Bibby, C. J., N. J. Collar, M. J. Crosby, M. F. Heath, C. R. Imboden, T. H. Johnson, A. J. Lang, A. J. Stattersfield, and S. J. Thirgood. 1993. *Putting diversity on the map: Priority areas for global conservation.* Cambridge: Burlington Press.

Brown, J. H. 1995. *Macroecology.* Chicago: University of Chicago Press.

Brown, J. H., and A. C. Gipson. 1983. *Biogeography.* St. Louis: C. V. Mosby Co.

Collar, N. J., and S. N. Stuart. 1988. *Key forest for threatened birds in Africa,* Monograph no. 3. Cambridge: International Council for Bird Preservation.

Darwin, C. 1898. *Journal of researches into the natural history and geology of the countries visited during the voyage of the H.M.S. Beagle round the world.* New York: Appleton.

Davis, S., and J. Botkin. 1994. The coming of knowledge-based business. *Harvard Business Review* (September–October): 165–168.

Dinerstein, E., and E. D. Wikramanayake. 1993. Beyond "hotspots": How to prioritize investments to conserve biodiversity in the Indo-Pacific Region. *Conservation Biology* 7: 53–65.

Dobson, A. P., J. P. Rodriguez, and D. S. Wilcove. 1997. Geographic distribution of endangered species in the United States. *Science* 275: 550–555.

Ehrlich, P. R. 1988. Tomorrow's world: Why saving biodiversity is today's priority. *World Birdwatch* 10: 6–8.

Howard, A., and J. Magretta. 1995. Surviving success: An interview with Nature Conservancy's John Sawhill. *Harvard Business Review* (September–October): 109–118.

Huston, M. A. 1994. *Biological diversity. The coexistence of species on changing landscapes.* Cambridge: Cambridge University Press.

Knopf, F. L. 1992. Faunal mixing, faunal integrity, and the biopolitical template for diversity conservation. *Transactions of the North American Wildlife and Natural Resources Conference* 57: 330–342.

Kruckeberg, A. R., and D. Rabinowitz. 1985. Biological aspects of endemism in higher plants. *Annual Review of Ecology and Systematics* 16: 447–479.

MacArthur, R. 1972. *Geographical ecology: Patterns in the distribution of species.* Princeton: Princeton University Press.

MacArthur, R., and E. O. Wilson. 1967. *The theory of island biogeography.* Princeton: Princeton University Press.

Mares, M. A. 1992. Neotropical mammals and the myth of Amazonian biodiversity. *Science* 255: 976–980.

Mintzberg, H. 1994. The fall and rise of strategic planning. *Harvard Business Review* (January–February): 107–113.

Myers, N. 1988. Threatened biotas: "Hotspots" in tropical forests. *Environmentalist* 8: 187–208.

Pielou, E. C. 1975. *Ecological diversity.* New York: Wiley.

Pimm, S. L., and J. L. Gittleman. 1992. Biological diversity: Where is it? *Science* 255: 940.

Poole, A., and F. B. Gill. 1998. *The birds of North America.* Philadelphia: Smith-Edwards-Dunlap Company.

Posner, B. G., L. R. Rothstein, C. Horner, G. H. B. Ross, and D. F. Kettl. 1994. Reinventing the business of government: An interview with catalyst David Osborne. *Harvard Business Review* (May–June): 132–144.

Pulliam, H. R., and B. Babbitt. 1997. Science and the protection of endangered species. *Science* 275: 499–500.

Samson, F. B., and F. L. Knopf. 1982. In search of a diversity ethic for wildlife management. *Transactions of the North American Wildlife and Natural Resources Conference* 47: 421–431.

_____. 1994. Prairie conservation in North America. *BioScience* 44: 418–421.

_____. 1996. Putting "ecosystem" into resource management. *Journal of Soil and Water Conservation* 51: 288–291.

Sisk, T. D., A. E. Lavner, K. R. Switky, and P. R. Ehrlich. 1994. Identifying extinction threats. *BioScience* 44: 592–604.

Vane-Wright, R. I., et al. What to protect: Systematics and the agony of choice. *Biological Conservation* 55: 235–254.

Wilcove, D. S., and R. B. Blair. 1995. The ecosystem management band-wagon. *Trends in Ecology and Evolution* 10: 345.

Wilcove, D. S., and F. B. Samson. 1987. Innovative wildlife management: Listening to Leopold. *Transactions of the North American Wildlife and Natural Resources Conference* 52: 321–329.

Wilson, E. O. 1994. Foreword to *Putting diversity on the map: Priority areas for global conservation*, ed. C. J. Bibby, N. J. Collar, M. J. Cosby, M. F. Heath, A. J. Stattersfield, and J. J. Thirgood. Cambridge: Burlington Press.

Wulff, E. V. 1943. *An introduction to historical plant geography*. Waltham, Mass.: Chronica Botanica.

Chapter 11

Arcese, P., and A. R. E. Sinclair. 1997. The role of protected areas as eco-logical baselines. *Journal of Wildlife Management* 61: 587–602.

Armstrong, D. P., T. Soderquist, and R. Southgate. 1994. Designing exper-imental reintroductions as experiments. In *Reintroduction biology of Australian and New Zealand fauna*, ed. M. Serena, 27–29. Chipping Norton, United Kingdom: Surrey Beatty and Sons.

Baskin, Y. 1994. Ecologists dare to ask: How much diversity is enough? *Science* 264: 202–203.

Boyce, M. S. 1993. Population viability analysis: Adaptive management for threatened and endangered species. *Transactions of the North American Wildlife Natural Resources Conference* 58: 520–527.

Brace, R. K., R. S. Pospahala, and R. L. Jessen. 1987. Background and objectives on stabilized duck hunting regulations: Canadian and U. S. per-spectives. *Transactions of the North American Wildlife Natural Resources Conference* 52: 177–185.

Capen, D. E. 1989. Political unrest, progressive research, and professional education. Personal reflections. *Wildlife Society Bulletin* 17: 335–337.

Clark, R. O., and A. W. Diamond. 1993. Restoring upland habitats in the Canadian prairies: Lost opportunity or management by design? *Transactions of the North American Wildlife Natural Resources Conference* 58: 551–564.

Clark, T. W. 1992. Practicing natural resource management with a policy orientation. *Environmental Management* 16: 423–433.

Conroy, M. J., and B. R. Noon. 1996. Mapping species richness for conser-vation of biological diversity: Conceptual and methodological issues. *Ecological Applications* 6: 763–773.

Glenn, S. M., and T. D. Nudds. 1989. Insular biogeography of mammals in Canadian parks. *Journal of Biogeography* 16: 261–268.

Grime, J. P. 1997. Biodiversity and ecosystem function: The debate deep-ens. *Science* 277: 1260–1261.

Gurd, D. B. 1996. Avian occurrence in Ontario boreal forests subjected to disturbance from timber harvest and fire. M.Sc. thesis, University of Guelph.

Haila, Y. 1997. A "natural" benchmark for ecosystem function. *Conservation Biology* 11: 300–301.

Hunter, M., Jr. 1996. Benchmarks for managing ecosystems: Are human activities natural? *Conservation Biology* 10: 695–697.

Hurlburt, S. H. 1984. Pseudoreplication and the design of ecological field experiments. *Ecological Monographs* 54: 187–211.

Johnson, F. A., C. T. Moore, W. L. Kendall, J. A. Dubovsky, D. F. Caithamer, J. R. Kelley, and B. K. Williams. 1997. Uncertainty and the management of mallard harvests. *Wildlife Management* 61: 202–216.

Johnson, K. H., K. A. Vogt, H. J. Clark, O. J. Schmitz, and D. J. Vogt. 1996. Biodiversity and the productivity and stability of ecosystems. *Trends in Ecology and Evolution* 11: 372–377.

Johnson, S. P., ed. 1993. *The Earth Summit: The United Nations Conference on Environment and Development.* Bk, 1. Norwell, Mass.: Kluwer Academic Publishers.

Karr, J. R. 1981. Assessment of biotic integrity using fish communities. *Fisheries* 6: 21–27.

_____. 1991. Biological integrity: A long neglected aspect of water resource management. *Ecological Applications* 1: 66–84.

Kurzejeski, E. W., R. L. Clawson, R. B. Renken, S. L. Sheriff, L. D. Vangilder, and Faaborg. 1993. Experimental evaluation of forest management: The Missouri ecosystem project. *Transactions of the North American Wildlife Natural Resources Conference* 58: 599–609.

Lancia, R. A., C. E. Braun, M. W. Collopy, R. D. Dueser, J. G. Kie, C. J. Martinka, J. D. Nichols, T. D. Nudds, W. R. Porath, and N. G. Tilghman. 1996. ARM! For the future: Adaptive resource management in the wildlife profession. *Wildlife Society Bulletin* 24: 436–442.

Lancia, R. A., T. D. Nudds, and M. L. Morrison. 1993. Adaptive resource management: Policy as hypothesis, management by experiment. Opening comments: Slaying slippery shibboleths. *Transactions of the North American Wildlife Natural Resources Conference* 58: 505–508.

Lubchenco, J., A. M. Olson, L. B. Brubaker, S. R. Carpenter, M. M. Holland, S. P. Hubbell, S. A. Levin, J. A. MacMahon, D. A. Matson, J. H. Melillo, H. A. Mooney, C. A. Peterson, H. R. Pulliam, L. A. Real, D. J. Regal, and P. G. Risser. 1991. The sustainable biosphere initiative: An ecological research agenda. *Ecology* 72: 371–412.

Ludwig, D., R. Hilborn, and C. Walters. 1993. Uncertainty, resource exploitation and conservation: Lessons from history. *Science* 260: 17, 36.

Macnab, J. 1983. Wildlife management as scientific experimentation. *Wildlife Society Bulletin* 11: 397–401.

McLaughlin, C. G. 1997. Spatial scale and variation in species richness: Patterns, parameters, and predictions. M.Sc. thesis, University of Guelph.

Moffat, A. S. 1996. Biodiversity is a boon to ecosystems, not species. *Science* 271: 1497.

Moore, J. A. 1985. Science as a way of knowing. *American Zoology* 25: 1–155.

Morrison, M. L. 1986. *Bird populations as indicators of environmental change*, vol. 3 of *Current ornithology*, ed. R. F. Johnston, 429–451. New York: Plenum Press.

Nichols, J. D. 1991. Science, population ecology, and the management of the American black duck. *Journal of Wildlife Management* 55: 790–890.

Nudds, T. D. 1979. Theory in wildlife conservation and management. *Transactions of the North American Wildlife Natural Resources Conference* 44: 277–288.

_____. 1998. Harvest effects on biodiversity. In *Status and trends of the nation's biological resources*, ed. M. Mac, P. A. Opler, C. E. Puckett Haecker, and P. D. Doran, 167–177. Washington, D.C.: USDI, U.S. Geological Survey.

Nudds, T. D., J. P. Bogart, D. M. Britton, H. A. Hager, A. L. A. Middleton, D. N. Poffer, D. P. Tate, K. E. Weistead, S. A. Marshall, C. M. Buddle, T. Woodcock, D. W. Larson, and E. Whelpdale. 1996. *Species-area relations of woody plants, vertebrates and selected invertebrates on islands of Georgian Bay: The role of national parks in evaluating null models of biodiversity.* Final report on file. Tobermory, Ontario: Parks Canada, c/o Fathom Five National Marine Park.

Nudds, T. D., and R. G. Clark. 1993. Landscape ecology, adaptive resource management and the North American Waterfowl Management Plan. In *Third Prairie Conference and Endangered Species Workshop*, ed. G. L. Holroyd, H. L. Dickson, M. Regnier, and H. C. Smith, 180–190. Provincial Mus. Alberta Nat. Hist. Occas. Pap. No. 19.

Nudds, T. D., and C. O. McLaughlin. 1997. *Submission to the Parks Canada Agency Constituent Group Consultation Panel.* Honey Harbour, Ontario: Parks Canada, Georgian Bay Islands National Park.

Nudds, T. D., and M. L. Morrison. 1991. Ten years after "reliable knowledge": Are we gaining? *Journal of Wildlife Management* 55: 757–760.

Oliver, C. H., B. J. Shuter, and C. K. Minns. 1995. Toward a definition of conservation principles for fisheries management. *Canadian Journal of Fisheries and Aquatic Sciences* 52: 1584–1594.

Ontario Environmental Assessment Board. 1994. *Reasons for decision and decision: Class environmental assessment by the Ministry of Natural Resources for timber management on Crown lands in Ontario.* Toronto: Environmental Assessment Board.

Ontario Forest Policy Panel. 1993. *Diversity: Forests, people, communities. Proposed comprehensive forest policy framework for Ontario.* Toronto: Queen's Printer.

Revised Statutes of Ontario. 1994. *An act to revise the Crown Timber Act to provide for the sustainability of Crown forests in Ontario. Legislative Assembly of Ontario.* Toronto: Queen's Printer.

Romesburg, H. C. 1981. Wildlfie science: Gaining reliable knowledge. *Journal of Wildlife Management* 45: 293–313.

Sarrazin, F., and R. Barbault. 1996. Reintroduction: Challenges and lessons for basic ecology. *Trends in Ecology and Evolution* 11: 474–478.

Schmiegelow, F. K. A. 1992. Use of atlas data to test appropriate hypotheses about faunal collapse. In *Landscape approaches to wildlife and ecosystem management*, ed. G. B. Ingramand M. R. Moss, 67–74. Montreal: Polyscience Publications.

Schmiegelow, F. K. A., and S. J. Hannon. 1993. Adaptive management, adaptive science and the effects of forest fragmentation on boreal birds in northern Alberta. *Transactions of the North American Wildlife Natural Resources Conference* 58: 584–598.

Schmiegelow, F. K. A., C. S. Machtans, and S. J. Hannon. 1997. Are boreal birds resilient to fragmentation? An experimental study of short-term community responses. *Ecology* 78: 1914–1932.

Society of American Foresters. 1993. Sustaining long-term forest health and productivity. *Journal of Forestry* 91: 32–35.

Stelfox, J. B., ed. 1995. *Relationships between stand age, stand structure and biodiversity in aspen mixed wood forests in Alberta*. Vegreville, Alberta: Alberta Environmental Centre; Edmonton, Alberta: Canadian Forest Service.

Stohlgren, T. J., J. F. Quinn, M. Ruggieero, and G. S. Waggoner. 1995. Status of biotic inventories in U.S. national parks. *Biological Conservation* 71: 97–106.

U.S. Environmental Protection Agency (EPA). 1993. *Announcement. Request for applications. EMAP-0l-93*. Indicators of ecosystem stress. Washington, D.C.: Environmental Protection Agency.

Walters, C. J. 1986. *Adaptive management of renewable resources*. New York: Macmillan.

_____. 1992. Perspectives on adaptive policy design in fisheries management. In *Applied population biology*, ed. S. K. Jam and L. W. Botsford, 249–262. Boston, Mass.: Academic Publishers.

Williams, B. K., and F. A. Johnson. 1995. Adaptive management and the regulation of waterfowl harvests. *Wildlife Society Bulletin* 23: 430–436.

Williams, B. K., F. A. Johnson, and K. Wilkins. 1996. Uncertainty and the adaptive management of waterfowl harvests. *Journal of Wildlife Management* 60: 223–232.

Chapter 12

Chadwick, D. H. 1990. *The biodiversity challenge*. Washington, D.C.: Defenders of Wildlife.

Christensen, N. L., A. M. Bartuska, J. H. Brown, S. Carpenter, C. D'Antonio, R. Francis, J. F. Franklin, J. A. MacMahon, R. F. Noss, D. J. Parsons, C. H. Peterson, M. G. Turner, and R. G. Woodmansee. 1996. The report of the Ecological Society of America committee on the scientific basis for ecosystem management. *Ecological Applications* 6(3): 665–691.

Committee on Environment and Natural Resources. 1994. *Biodiversity and ecosystem dynamics: A strategy and implementation plan*. Washington,

D.C.: Committee on Environment and Natural Resources, Office of Science and Technology Policy.

Ehrenfeld, D. M. 1991. The management of diversity: A conservation paradox. In *Ecology, economics, ethics: The broken circle*, ed. F. H. Bormann and S. R. Kellert, 26–39. New Haven: Yale University Press.

Ehrlich, P. A., and E. O. Wilson. 1991. Biodiversity studies: Science and policy. *Science* 253: 758–762.

Gordon, J. C., and J. Lyons. 1997. The emerging role of science and scientists in ecosystem management. In *Creating a forestry for the 21st century: The science of ecosystem management*, ed. K. A. Kohm and J. F. Franklin, 447–453. Washington, D.C.: Island Press.

Grumbine, R. E. 1994. What is ecosystem management? *Conservation Biology* 8(1): 27–38.

Haufler, J. B. 1995. Forest industry partnerships for ecosystem management. *Transactions of North American Wildlife and Natural Resources Conference* 60: 422–432.

Interagency Ecosystem Management Task Force. 1995. *The ecosystem approach: Healthy ecosystems and sustainable economies.* Vol. 1. Washington, D.C.: Interagency Ecosystem Management Task Force.

Lamm, R. D. 1997. *U.S. population demographics, sustainable development, and science.* A presentation to the U.S. Geological Survey's Geologic Division Science Strategy Team, Denver.

Lanica, R. A., T. D. Nudds, and M. L. Morrison. 1993. Adaptive resource management: Policy as hypothesis, management by experiment. Opening comments: Slaying slippery shibboleths. *Transactions of North American Wildlife and Natural Resources Conference* 58: 505–508.

Lubchenco, J., A. M. Olson, L. B. Brubaker, S. R. Carpenter, M. M. Holland, S. P. Hubbell, S. A. Levin, J. A. MacMahon, P. A. Matson, J. M.Melillo, H. A. Mooney, C. H. Peterson, H. R. Pulliam, L. A. Real, P. J. Regal, and P. G. Kisser. 1991. The sustainable biosphere initiative: An ecological agenda. *Ecology* 72(2): 371–412.

Mares, M. A. 1992. Neotropical mammals and the myth of Amazonian biodiversity. *Science* 255: 976–980.

Meffe, G. K., and C. R. Carroll. 1994. *Principles of conservation biology.* Sunderland, Mass.: Sinauer Associates.

McNeely, J. A., K. R. Miller, W. V. Reid, R. A. Mittermeier, and T. B. Werner. 1990. *Conserving the world's biological diversity.* Gland, Switzerland: International Union for Conservation of Nature and Natural Resources; Washington, D.C.: World Resources Institute, Conservation International, World Wildlife Fund—U.S., and the World Bank.

Miller, K. R., J. Furtado, C. DeKlemm, J. A. Neely, N. Myers, M. E. Soulé, and M. C. Trexler. 1985. Issues on the preservation of biological diversity. In *The global possible: Resources development and the new century*, ed. R. Repetto, 337–362. New Haven: Yale University Press.

Morrisette, P. M. 1992. Congress on renewable natural resources: Critical issues and concepts for the twenty-first century. *Renewable Resources Journal* 10(3): 3–31.

National Research Council. 1993. *A biological survey for the nation.* Washington, D.C.: National Academy Press.

National Science and Technology Council. 1995. *Preparing for the future through science and technology: an agenda for environmental and natural resources research.* Washington, D.C.: National Science and Technology Council.

Ohlendorf, H. M., R. L. Hothem, C. M. Bunck, and K. C. Marois. 1990. Bioaccumulation of selenium in birds at Kesterson Reservoir, California. *Archives of Environmental Contamination and Toxicology* 19: 495–507.

Ryan, J. C. 1992. Conserving biological diversity. In *State of the world 1992*, ed. L. R. Brown, C. Flavin, S. Postel, and L. Starke, 9–26. New York: W. W. Norton.

Salwasser, H. 1991. New perspectives for sustaining diversity in the U.S. national forest ecosystems. *Conservation Biology* 5(4): 567–569.

Sampson, F. B., and F. L. Knopf. 1993. Managing biological diversity. *Wildlife Society Bulletin* 21: 509–514.

Schonewald-Cox, C. 1994. Protection of biological diversity: Missing connection between science and management. In *Biological diversity: Problems and challenges*, ed. S. S. Majumdar, F. J. Brenner, J. E. Lovich, J. F. Schalles, and E. W. Miller, 171–183. Pittsburgh: Pennsylvania Academy of Science Press.

Soulé, M. E., and M. A. Kohm. 1989. *Research priorities for conservation biology.* Washington, D.C.: Island Press.

Strauss, W., and N. Howe. 1997. *The fourth turning.* New York: Broadway Books.

Trauger, D. L., and R. J. Hall. 1992. The challenge of biological diversity: Professional responsibilities, capabilities, and realities. *Transactions of North American Wildlife and Natural Resources Conference* 57: 20–36

Trauger, D. L., W. C. Tilt, and C. B. Boyd Hatcher. 1995. Partnerships: Innovative strategies for wildlife conservation. *Wildlife Society Bulletin* 23(1): 114–119.

U.S. Government General Accounting Office. 1994. *Ecosystem management: Additional actions needed to adequately test a promising approach.* Washington, D.C.: U.S. Government Accounting Office.

Van Home, B., and J. A. Wiens. 1991. Forest bird habitat suitability models and the development of general habitat models. *Fish and Wildlife Research* (U.S. Fish and Wildlife Service) 8.

Walters, C. 1986. *Adaptive management of natural resources.* New York: Macmillan.

Weston, D. 1992. The biodiversity crisis. A challenge for biology. *Oikos* 63: 29–38.

Wilson, E. O. 1988. *Biodiversity.* Washington, D.C.: National Academy Press.

_____. 1992. *The diversity of life.* New York: W. W. Norton.

Wolf, E. C. 1987. On the brink of extinction: Conserving the diversity of life. *Worldwatch Paper* 78: 1–53.

Chapter 13

Allen, A. W. 1993. Wildlife habitat criteria in relation to future use of CRP lands. *Proceedings of the Great Plains Agricultural Conference* 41–88. Rapid City, S.D.

Bain, M. B. 1993. Assessing impacts of introduced aquatic species: Grass carp in large systems. *Environmental Management* 17: 211–224.

Best, L. B., H. Campa III, K. E. Kemp, R. J. Robel, M. R. Ryan, J. A. Savidge, H. P. Weeks, Jr., and S. R. Winterstein. 1997. Bird abundance and nesting in CRP fields and cropland in the midwest: A regional approach. *Wildlife Society Bulletin* 25: 864–877.

Bock, C. E., J. H. Bock, K. L. Jepson, and J. C. Ortega. 1986. Ecological effects of planting African lovegrasses in Arizona. *National Geographic Research* 2: 456–463.

Coon, T. G. In press. Great Lakes ichthyofauna. In *Great Lakes fisheries policy and management: A binational perspective*, ed. W. W. Taylor and C. P. Fererri. East Lansing: Michigan State University Press.

Crossman, E. J. 1981. Introduced freshwater fishes: A review of the North American perspective with emphasis on Canada. *Canadian Journal of Fisheries and Aquatic Science* 48: 46–57.

D'Antonio, C. M., and P. M. Vitousek. 1992. Biological invasions by exotic grasses, the grass/fire cycle. *Annual Review of Ecology and Systematics* 23: 63–88.

Delisle, J. M., and J. A. Savidge. 1997. Avian use and vegetation characteristics of Conservation Reserve Program fields. *Journal of Wildlife Management* 61: 318–325.

Ehrenfeld, D. W. 1970. *Biological conservation*. New York: Holt, Rinehart, and Winston.

Gamradt, S. C., and L. B. Katz. 1996. Effects of introduced crayfish and mosquito fish on California newts. *Conservation Biology* 10: 1155–1162.

Hanaburgh, C. 1995. Wildlife use of native and introduced grasslands in Michigan. M.S. thesis, Michigan State University.

Hardin, S., R. Land, M. Spelman, and G. Morse. 1984. Food items of grass carp, American coots, and ring-necked ducks from a central Florida lake. *Proceedings of the Annual Conference Southeastern Association of Fish and Wildlife Agencies* 38: 313–318.

Herkert, J. R., D. W. Sample, and R. E. Warner. 1996. Management of midwestern grassland landscapes for the conservation of migratory birds. *Management of midwestern landscapes for the conservation of migratory birds*, ed. F. R. Thompson, 89–116. USDA Forest Service, North-Central Experiment Station, General Technical Report NC-187.

Holm E., and G. A. Coker. 1981. First Canadian records of the ghost shiner (*Notropis buchanani*) and the orange spotted sunfish (*Lepomis humilis*). *Canadian Field Naturalist* 95: 210–211.

Hubbs, C. L., and K. F. Lagler. 1947. *Fishes of the Great Lakes region*. Bulletin no. 26, 186 pp. Bloomfield Hills, Mich.: Cranbrook Institute of Science.

Jenkins, P. 1996. Harmful exotics in the United States. *Biodiversity and law*, ed. W. J. Snape, 105–119. Washington, D.C.: Island Press.

Lawrie, A. H. 1970. The sea lamprey in the Great Lakes. *Transactions of the American Fisheries Society* 99: 766–775.

Leege, L. M. 1997. The ecological impact of Austrian pine (*Pinus nigra*) on sand dunes of Lake Michigan: An introduced species becomes an invader. Ph.D. dissertation, Michigan State University.

Lever, L. M. 1996. *Naturalized fishes of the world*. San Diego: Academic Press.

Lindell, C. 1996. Patterns of nest usupation: When should species converge on nest niches? *Condor* 98: 464–473.

Mauchamp, A. 1997. Threats from alien plant species in the Galapagos Islands. *Conservation Biology* 11: 260–263.

McKnight, S. K., and G. R. Hepp. 1995. Potential effect of grass carp herbivory on waterfowl foods. *Journal of Wildlife Management* 59: 720–727.

Meffe, G. K., and C. R. Carroll. 1997. *Principles of conservation biology*, 2d ed. Sunderland, Mass.: Sinauer Associates.

Moulton, M. P., and J. Sanderson. 1997. *Wildlife issues in a changing world*. Delray Beach, Fla.: St. Lucie Press.

Nalepa, T. F., and D. W. Schloesser, eds. 1993. *Zebra mussels. Biology, impacts, and control*. Boca Raton, Fla.: CRC Press.

Office of Technology Assessment (OTA). 1993. *Harmful non-indigenous species in the United States*. Washington, D.C.: U.S. Government Printing Office, OTA-F-565.

Opler, P. A. 1978. *Insects of American chestnut: Possible importance and conservation concern. The American chestnut symposium*. Morgantown: West Virginia University Press.

Osborn, T., and R. Heimlich. 1994. Changes ahead for the Conservation Reserve Program. *Agricultural Outlook* 209: 26–30.

Pedersen, E. K., W. E. Grant, and M. T. Longnecker. 1996. Effects of red imported fire ants on newly hatched northern bobwhite. *Journal of Wildlife Management* 60: 164–169.

Pine, R. T., and W. J. Anderson. 1991. Plant preference of triploid grass carp. *Journal of Aquatic Plant Management* 29: 80–82.

Pogue, D. W., and G. D. Schnell. 1994. Habitat characterization of secondary cavity-nesting birds in Oklahoma. *Wilson Bulletin* 106: 203–226.

Reichard, S. H., and C. W. Hamilton. 1997. Predicting invasions of woody plants introduced into North America. *Conservation Biology* 11: 193–203.

Rodda, G. H., and T. H. Fritts. 1992. The impact of the introduction of the brown tree snake, *Boiga irregularis*, on Guam's lizards. *Journal of herpetology* 26: 166–174.

Ruesink, J. L., I. Parker, M. Groom, and P. Kareiva. 1995. Reducing the risks of nonindigenous species introductions. *BioScience* 45: 465–477.

Sauer, J. R., and S. Droege. 1990. Recent population trends of the eastern bluebird. *Wilson Bulletin* 102: 239–252.

Savidge, J. A. 1987. Extinction of an island avifauna by an introduced snake. *Ecology* 68: 660–668.

Schoulties, C. L. 1991. *Pathways and consequences of the introduction of non-indigenous plant pathogens in the United States.* Washington, D.C.: Office of Technology Assessment.

Schultz, E. E. 1960. Establishment and early dispersal of a loach, *Misgurnus anguillicaudatus*, in Michigan. *Transactions of the American Fisheries Society* 89: 376–377.

Chapter 14

Alverson, W. S., W. Kuhlmann, and D. M. Waller. 1994. *Wild forests: Conservation biology and public policy.* Washington, D.C.: Island Press.

Blockstein, D. E. 1995. A strategic approach for biodiversity conservation. *Wildlife Society Bulletin* 23: 365–369.

Botkin, D. B. 1990. *Discordant harmonies: A new ecology for the twenty-first century.* New York: Oxford University Press.

Camp, A., C. Oliver, P. Hessburg, and R. Everett. 1997. Predicting late successional fire refugia predating European settlement in the Wenatchee Mountains. *Forest Ecology and Management* 95: 63–77.

Davis, J. 1994. WE role in the wildlands. *Wild Earth, The Wildlands Project* (special issue): 9.

Davis, M. B., S. Sugita, R. R. Calcote, J. B. Ferrari, and L. E. Frelich. 1994. Historical development of alternative communities in a hemlock-hardwood forest in northern Michigan, USA. In *Large-scale ecology and conservation biology*, ed. P. J. Edwards, R. M. May, and N. R. Webb, 19–39. Oxford, England: Blackwell Science.

DellaSala, D. A., D. M. Olson, S. E. Barth, S. L. Crane, and S. A. Primm. 1995. Forest health: Moving beyond rhetoric to restore healthy landscapes in the inland Northwest. *Wildlife Society Bulletin* 23: 346–356.

DellaSala, D. A., J. R. Strittholt, R. F. Noss, and D. M. Olson. 1996. A critical role for core reserves in managing inland northwest landscapes for natural resources and biodiversity. *Wildlife Society Bulletin* 24: 209–221.

Ecomap. 1993. *National hierarchical framework of ecological units.* Washington, D.C.: USDA Forest Service.

Everett, R. L., and J. Lehmkuhl. 1996. An emphasis-use approach to conserving biodiversity. *Wildlife Society Bulletin* 24: 192–199.

Harris, L. D. 1984. *The fragmented forest.* Chicago: University of Chicago Press.

Haufler, J. B., C. A. Mehl, and G. J. Roloff. 1996. Using a coarse-filter approach with species assessment for ecosystem management. *Wildlife Society Bulletin* 24: 200–208.

Hunter, M. L. 1990. *Wildlife, forests, and forestry. Principles of managing forests for biodiversity.* Englewood Cliffs, N.J.: Prentice Hall.

Jones, S. M., and F. T. Lloyd. 1993. Landscape ecosystem classification: The first step toward ecocystem management in the southeastern United States. In *Defining sustainable forestry*, ed. G. H. Aplet, N. Johnson, J. T. Olson, and V. A. Sample, 181–210. Washington, D.C.: Island Press.

Kaufmann, M. R., R. T. Graham, D. A. Boyce, Jr., W. H. Moir, L. Perry, R.

T. Reynolds, R. L. Bassett, P. Mehlhop, C. B. Edminster, W. M. Block, and P. S. Corn. 1994. *An ecological basis for ecosystem management.* USDA Forest Service, General Technical Report RM-246.

Kimmins, H. 1992. *Balancing act: Environmental issues in forestry.* Vancouver: University of British Columbia Press.

Marcot, B. G., M. J. Wisdon, H. W. Li, and G. C. Castillo. 1994. *Managing for featured, threatened, endangered, and sensitive species and unique habitats for ecosystem sustainability.* USDA Forest Service, General Technical Report PNW-GTR-329.

McMahon, J. 1993. *An ecological reserves system for Maine: Benchmarks in a changing landscape.* Bangor: Natural Resources Policy Division, Maine State Planning Office.

Morgan, P., G. H. Aplet, J. B. Haufler, H. C. Humphries, M. M. Moore, and W. D. Wilson. 1994. Historical range of variability: A useful tool for evaluating ecosystem change. *Journal of Sustainable Forestry* 2: 87–112.

Morrison, M. L., B. G. Marcot, and R. W. Mannan. 1992. *Wildlife-habitat relationships: Concepts and applications.* Madison: University of Wisconsin Press.

Noss. R. F. 1990. What can wilderness do for biodiversity? In *Preparing to manage wilderness in the 21st century*, comp. P. C. Reed, 49–61. USDA Forest Service, General Technical Report SE-66.

_____. 1994. The wildlands project: Land conservation strategy. *Wild Earth, The Wildlands Project* (special issue): 10–25.

Noss, R. F., and A. Y. Cooperrider. 1994. *Saving nature's legacy: Protecting and restoring biodiversity.* Washington, D.C.: Island Press.

O'Hara, K. L., P. A. Latham, P. Hessburg, and B. G. Smith. 1996. A structural classification for inland northwest forest vegetation. *Western Journal of Applied Forestry* 11(3): 97–102.

Oliver, C. D. 1992. A landscape approach: Achieving and maintaining biodiversity and economic productivity. *Journal of Forestry* 90: 20–25.

Oliver, C. D., A. Osawa, and A. Camp. 1997. Forest dynamics and resulting animal and plant population changes at the stand and landscape levels. *Journal of Sustainable Forestry* 6(3/4): 281–312.

Pfister, R. D. 1993. The need and potential for ecosystem management in forests of the inland west. In *Defining sustainable forestry*, ed. G. H. Aplet, N. Johnson, J. T. Olson, and V. A. Sample, 217–239. Washington, D.C.: Island Press.

Risbrudt, C. 1992. Sustaining ecological systems in the Northern Region. In *Taking an ecological approach to management*, 27–39. USDA Forest Service, Watershed and Air Management, WO-WSA-3.

Roloff, G. J. 1994. *Using an ecological classification system and wildlife habitat models in forest planning.* Ph.D. dissertation, Michigan State University.

Roloff, G. J., and J. B. Haufler. 1997. Establishing population viability planning objectives based on habitat potentials. *Wildlife Society Bulletin* 25: 895–904.

Salwasser, H., D. W. MacCleery, and T. A. Snellgrove. 1993. An ecosystem perspective on sustainable forestry and new directions for the U.S. National Forest System. In *Defining sustainable forestry*, ed. G. H. Aplet, N. Johnson, J. T. Olson, and V. A. Sample, 44–89. Washington, D.C.: Island Press.

Steele, R., R. D. Pfister, R. A. Ryker, and J. A. Kittams. 1981. *Forest habitat types of Central Idaho*. USDA Forest Service, General Technical Report INT-114.

Thomas, J. W., M. G. Raphael, and R. G. Anthony. 1993. *Viability assessments and management considerations for species associated with late-successional and old-growth forests of the Pacific Northwest*. Washington, D.C.: USDA Forest Service.

Chapter 15

Bailey, R. G. 1995. *Descriptions of the ecoregions of the United States*. USDA Forest Service, Misc. Publ. 1391.

_____. 1996. *Ecosystem geography*. New York: Springer-Verlag.

Daubenmire, R. F. 1968. *Plant communities: A textbook of plant synecology*. New York: Harper and Row.

Ecomap. 1993. *National hierarchical framework of ecological units*. Washington D.C.: USDA Forest Service.

Gregg, W. P., Jr. 1994. Developing landscape-scale information to meet ecological, economic, and social needs. In *Remote sensing and GIS in ecosystem management*, ed. V. A. Sample, 13–17.

Haufler, J. B. 1994. An ecological framework for forest planning for forest health. *Journal of Sustainable Forestry* 2: 307–316.

_____. 1995. Forest industry partnerships for ecosystem management. *Transactions of the North American Wildlife and Natural Resources Conference* 60: 422–432.

Haufler, J. B., T. Crow, and D. Wilcove. In press. Scale considerations for ecosystem management. *Proceedings of the Ecological Stewardship Workshop*.

Haufler, J. B., C. A. Mehl, and G. J. Roloff. 1996. Using a coarse-filter approach with species assessment for ecosystem management. *Wildlife Society Bulletin* 24: 200–208.

Keystone Center, The. 1996. *The Keystone national policy dialogue on ecosystem management: Final report*. Keystone, Colo.: The Keystone Center.

Maxwell, J. R., C. J. Edwards, M. E. Jensen, S. J. Paustian, H. Parrott, and D. M. Hill. 1995. *A hierarchical framework of aquatic ecological units in North America (Nearctic zone)*. USDA Forest Service, General Technical Report NC-176.

Oliver, C. D. 1992. A landscape approach: Achieving and maintaining biodiversity and economic productivity. *Journal of Forestry* 90: 20–25.

Roloff, G. J., and J. B. Haufler. 1997. Establishing population viability planning objectives based on habitat potentials. *Wildlife Society Bulletin* 25: 895–904.

Russell, W. E., and J. K. Jordan. 1991. Ecological classification systems for classifying land capability in midwestern and northeastern U.S. national forests. *Proceedings of a symposium: Ecological land classification: Applications to identify the productive potential of southern forests*, ed. D. L. Mengel and D. T. Tew, 18–24. USDA Forest Service, General Technical Report SE-68.

Walters, C. J. 1986. *Adaptive management of renewable resources*. New York: Macmillan.

Chapter 16

Landres, P. B., J. Verner, and J. W. Thomas. 1988. Ecological uses of vertebrate indicator species: A critique. *Conservation Biology* 2: 316–328.

Niemi, G. J., J. M. Hanowski, A. R. Lima, T. Nicholls, and N. Weiland. 1997. A critical analysis on the use of indicator species in management. *Journal of Wildlife Management* 61: 1240–1252.

Roloff, G. J., and J. B. Haufler. 1993. Forest planning in Michigan using a GIS approach. *Proceedings of the International Union of Game Biologists: Forest and wildlife . . . towards the 21st century*, vol. 1, XXI Congress. Halifax, Canada, ed. I. D. Thompson, 364–369.

Runkle, J. R. 1985. Disturbance regimes in temperate forests. In *The ecology of natural disturbances and patch dynamics*, ed. S. T. A. Pickett and P. S. White, 17–34. San Diego: Academic Press.

Contributors

Dr. Gregory H. Aplet is forest ecologist in the Ecology and Economics Research Department of The Wilderness Society, Denver, Colorado.

Dr. Richard K. Baydack is associate director of the Natural Resources Institute at the University of Manitoba, Winnipeg, Canada. His research interests are wildlife habitat management, conservation of biodiversity, and adaptive resource management, and he teaches courses in wildlife management, research planning, and natural resource ecology.

Dr. Henry (Rique) Campa III is associate professor of wildlife ecology at Michigan State University, East Lansing, Michigan. His research interests are in the areas of wildlife–habitat relationships, herbivory, and wildlife nutrition.

Dr. Allen Y. Cooperrider is senior biologist at the U.S. Fish and Wildlife Service, Ukiah, California.

Patrick J. Crist is Western regional coordinator of the GAP Analysis Program in the Biological Resources Division of the U.S. Geological Survey, Moscow, Idaho.

Dr. Blair Csuti is adjunct associate professor in the Department of Fish and Wildlife Resources at the University of Idaho, Moscow, Idaho.

Steven Day is the founder of LEGACY—The Landscape Connection, Leggett, California.

Dr. Richard L. Everett is the retired science team leader at the USDA Forest Service in the Pacific Northwest Research Station, Wenatchee, Washington.

Christine Hanaburgh is a Ph.D. candidate in the Department of Fisheries and Wildlife at Michigan State University.

Dr. Jonathan B. Haufler is manager of wildlife and ecology at the Boise Cascade Corporation, Boise, Idaho, where he is involved with ecosystem management initiatives at landscape levels. Prior to this position, he was professor of wildldife ecology at Michigan State University.

Curtice Jacoby is GIS analyst in the Spatial Analysis Lab at Humboldt State University, Arcata, California.

Michael D. Jennings is director of the GAP Analysis Program in the Biological Resources Division at the U.S. Geological Survey, Moscow, Idaho.

William S. Keeton is a doctoral student at the College of Forest Resources at the University of Washington, Seattle, Washington.

Brian J. Kernohan is project manager and wildlife biologist in the Minnesota Ecosystem Management Project at the Boise Cascade Corporation, International Falls, Minnesota.

Dr. Fritz L. Knopf is senior scientist at the Midcontinental Research Center in the Biological Resources Division of the U.S. Geological Survey, Fort Collins, Colorado.

Dr. John F. Lehmkuhl is science team leader and research wildlife biologist in the Pacific Northwest Research Station at the USDA Forest Service, Wenatchee, Washington.

Carolyn A. Mehl is president of Wildlife and Ecosystem Management Associates, Boise, Idaho, with prior experience with Boise Cascade Corporation, EMA Consultants, U.S. Fish and Wildlife Service, and the Army Corps of Engineers.

Dr. Jeffrey W. Nelson is director of operations of the Great Plains Regional Office at Ducks Unlimited Inc., Bismarck, North Dakota.

Dr. Terry G. Neraasen is managing director of the Alberta Conservation Association, Edmonton, Canada.

Dr. Thomas D. Nudds is professor of wildlife ecology in the Department of Zoology at the University of Guelph, Guelph, Canada.

Dr. Gary J. Roloff is wildlife management specialist with Boise Cascade Corporation, Boise, Idaho, with previous experience with the Missouri Department of Conservation.

Dr. Fred B. Samson is regional wildlife ecologist in the northern region of the U.S. Department of Agriculture, Forest Service division, Missoula, Montana.

Dr. J. Michael Scott is leader of the Biological Resources Division of the U.S. Geological Survey in the Idaho Cooperative Fish and Wildlife Unit, University of Idaho, Moscow, Idaho.

Dr. David L. Trauger is senior staff biologist and leader of the Fisheries and Wildlife Science Team in the Biological Resources Division of the U.S. Geological Survey, Reston, Virginia.

Dr. William A. Wall is staff specialist with Safari Club International, with previous experience with Potlach Corporation in Idaho, and with International Paper in Texas.

Dr. R. Gerald Wright is research scientist in the Biological Resources Division at the U.S. Geological Survey in the Idaho Cooperative Fish and Wildlife Research Unit, University of Idaho, Moscow, Idaho.

Index

Adaptation
    biodiversity and, 170
    genetic variability in, 13–14
Adaptive management, 179–193, 234
    application of, 247–248
    in intensively managed forests, 129,
        137–139
    uncertainty and, 197–198
Administrative boundaries, 22, 32, 88, 163,
    222
Africa, 13, 15
African clawed frogs, 204
African lovegrass, 208
Agelaius phoeniceus, 162
Agro-ecosystems, 60, 82
    Canadian Prairie Pothole Region,
        141–165
    grass planting in erodible cropland,
        211–212
    modified agricultural use in, 154–155
Alabama, exotic species in, 208
Alectoris chukar, 204
Alien species. *See* Exotic species
Alopecurus aequalis var. *sonomensis,* 95
Alpha diversity, 170
Amazon rainforests, 170–171
Ambystoma tigrinium, 162
American bitterns, 162
American chestnut, 209, 226
American golfinches, 162
American wigeons, 141, 142
Ammodramus bairdii, 162
Ammodramus henslowi, 211
Ammodramus leconteii, 161
Ammodramus savannarum, 162, 211
Amphibians, 60–61, 62, 139
    Canadian Prairie Pothole Region, 162
    exotic species and, 204, 206–207
Anas acuta, 141, 142
Anas americana, 141, 142
Anas clypeata, 141
Anas discors, 141

Anas platyrhynchos, 141, 149, 157–159
Anas strepera, 141
Anthus spragueii, 162
Antitrust issues, 234
Apis mellifera, 204
Aquatic communities, 29–30, 59, 95, 128,
    134
Aristotle, 7
Artemisia tridentata, 208
Arthropods, 172
Assessment, "inventory, monitor, and
    assess," 186–187. *See also* Species
    assessment
Atlasing, 138
Australia, 56, 68, 208, 209–210
Austrian pine, 205
Aythya affinis, 142
Aythya americana, 141
Aythya valisineria, 141

Baird's sparrows, 162
Baker's cypress, 62
Bald eagles, 203
Baseline data, 36, 148, 184. *See also*
    Historical range of variability
Bell's vireo, 68
Benchmark areas, 36, 148–152
Beta diversity, 170
Betula uber, 168
Biblical writings, 7
Biodiversity
    baseline data for. *See* Baseline data;
        Historical range of variability
    definitions, 1–6, 7, 169–174
    hierarchical, 170, 185–186, 237
    historical background, 7–9
    implementation of conservation process,
        233–249
    larger perspective for, 33–34
    limiting factors for, 241–242
    need for, 11–15
    quantitation of, 6–7

303